William
CONST
AND PC

John E.
CONST

William R. Park
CONSTRUCTION BIDDING FOR PROFIT

J. Stewart Stein
CONSTRUCTION GLOSSARY: AN ENCYCLOPEDIC REFERENCE AND MANUAL

James E. Clyde
CONSTRUCTION INSPECTION: A FIELD GUIDE TO PRACTICE

Harold J. Rosen and Philip M. Bennett
CONSTRUCTION MATERIALS EVALUATION AND SELECTION: A SYSTEMATIC APPROACH

C. R. Tumblin
CONSTRUCTION COST ESTIMATES

Harvey V. Debo and Leo Diamant
CONSTRUCTION SUPERINTENDENT'S JOB GUIDE

Oktay Ural, Editor
CONSTRUCTION OF LOWER-COST HOUSING

Robert M. Koerner and Joseph P. Welsh
CONSTRUCTION AND GEOTECHNICAL ENGINEERING USING SYNTHETIC FABRICS

Construction and Geotechnical Engineering Using Synthetic Fabrics

Construction and Geotechnical Engineering Using Synthetic Fabrics

Robert M. Koerner, Ph.D., P.E.

Professor of Civil Engineering
Drexel University
Philadelphia, PA

Joseph P. Welsh, P.E.

Vice President
Hayward Baker Company
Odenton, MD

A Wiley-Interscience Publication

JOHN WILEY & SONS, New York • Chichester • Brisbane • Toronto

Library of Congress Cataloging in Publication Data

Koerner, Robert M 1933–
Construction and geotechnical engineering using synthetic fabrics.

(Wiley series of practical construction guides)
"A Wiley-Interscience publication."
Includes bibliographical references and index.
1. Synthetic fabrics in building. 2. Textile fibers, Synthetic. I. Welsh, Joseph P., 1933– joint author. II. Title.
TA668.K63 624′.189′7 79-21733
ISBN 0-471-04776-7

Printed in the United States of America

10 9 8 7 6 5 4 3 2 1

Series Preface

The Wiley Series of Practical Construction guides provides the working constructor with up-to-date information that can help to increase the job profit margin. These guidebooks, which are scaled mainly for practice, but include the necessary theory and design, should aid a construction contractor in approaching work problems with more knowledgeable confidence. The guides should be useful also to engineers, architects, planners, specification writers, project managers, superintendents, materials and equipment manufacturers, and, the source of all these callings, instructors and their students.

Construction in the United States alone will reach $250 billion a year in the early 1980s. In all nations, the business of building will continue to grow at a phenomenal rate, because the population proliferation demands new living, working, and recreational facilities. This construction will have to be more substantial, thus demanding a more professional performance from the contractor. Before science and technology had seriously affected the ideas, job plans, financing, and erection of structures, most contractors developed their know-how by field trial-and-error. Wheels, small and large, were constantly being reinvented in all sectors, because there was no interchange of knowledge. The current complexity of construction, even in more rural areas, has revealed a clear need for more proficient, professional methods and tools in both practice and learning.

Because construction is highly competitive, some practical technology is necessarily proprietary. But most practical day-to-day problems are common to the whole construction industry. These are the subjects for the Wiley Practical Construction Guides.

M. D. Morris, P.E.

New York, New York

Preface

Over the very recent past (5 years, perhaps 10 years at the most) the construction industry has seen the birth, growth, and rapid expansion of synthetic fabric use. While the initial use of fabrics in construction was aimed at road reinforcement and more economic underdrain systems, the use of fabrics has spread to separation of materials, erosion control systems, and as flexible forms for sand, grout, and concrete. Furthermore, by rendering these synthetic fabrics (all are formed from man-made fibers) impermeable, a new group of construction uses has emerged. Included in this category are reservoir and tank liners, above ground uses such as air-supported and tension structures, water-filled structures, and even self-sustaining fabric structures.

In this book we deal with each of these topics, with the exception of reservoir and tank liners, a subject that has recently been published in this Wiley series under separate cover. Emphasis is placed on the below ground uses common to heavy construction (vis-à-vis building construction) and to geotechnical engineering.

This book should be of interest to contractors, consulting engineers, highway engineers, architects, site developers, and owners of sites with ground control problems. Additionally, it should be helpful to any persons—researchers, teachers, students, and so on—desiring an overview of the use of synthetic fabrics in construction.

As with many new ideas that are tried and proved to be successful, the fundamental theory and design methodology is usually lagging behind. This is the case with construction fabrics. Thus it is necessary, although unfortunate, that the thrust of the book is more towards an explanation of how fabrics function than why they function. In other words, use, rather than design, is emphasized. It is our hope that we have collected in one volume the materials necessary to understand what synthetic fabrics are, how they are currently evaluated, and how they are used in construction. Wherever available, performance data is also included.

It is hoped that this book will form the catalyst for subsequent work in the analysis and design of construction fabrics. The ultimate goal then becomes the development of realistic construction specifications that provide the owner with a safe installation and the contractor and manufacturer with an economic one.

ROBERT M. KOERNER, Ph.D., P.E.
JOSEPH P. WELSH, P.E.

Philadelphia, Pennsylvania
Odenton, Maryland
January 1980

Acknowledgments

To the manufacturers and suppliers of construction fabrics who have freely made available their technical information and experiences in this area we express our sincere thanks. The Wiley reviewers, Lee E. Murch of duPont and Richard G. Ahlvin of the Corps of Engineers, have made many helpful suggestions which we included.

We are particularly indebted to those members of research organizations and universities, and to other individuals who have taken the time and made the effort to publish their experiences in the open literature. These are the real stepping stones from which progress is made.

Contents

Conversion Table *xv*

1 *Overview and Background of Synthetic Fibers* *1*

1.1 Synthetic Fiber Statistics 1
1.2 Synthetic Fiber Production 4
1.2.1 The Fiber Base 4
1.2.2 The Spinneret and Related Processing 5
1.2.3 Subsequent Operations 5
1.2.4 Forms of Synthetic Fibers 7
1.3 Generic Names and Variants 8
1.4 Some Physical Properties of Synthetic Fibers 12
1.5 References 12

2 *Construction Fabrics* *13*

2.1 Elements of Fabric Construction 13
2.1.1 Woven Fabrics 13
2.1.2 Nonwoven Fabrics 16
2.2 Overview of Construction Fabric Uses 21
2.2.1 Separation 22
2.2.2 Reinforcement 23
2.2.3 Drainage 24
2.2.4 Erosion Control 24
2.2.5 Forms 25
2.2.6 Impermeable Fabrics 25
2.3 Fabric Properties of Importance in Construction Use 26
2.3.1 Tests for Physical Properties 26
2.3.2 Tests for Mechanical Properties 27
2.3.3 Tests for Hydraulic Properties 31

2.3.4 Tests for Endurance and Miscellaneous Properties 34

2.4 Overview of Construction Fabrics 36

2.5 Details of Construction Fabrics 36

2.5.1 Adva-Felt 38

2.5.2 Bay Mills 39

2.5.3 Bidim 39

2.5.4 Cerex 40

2.5.5 Cordura 40

2.5.6 Enkamat 40

2.5.7 Fibretex 44

2.5.8 Filter-X 44

2.5.9 Laurel Cloth 46

2.5.10 Mirafi 47

2.5.11 Monofelt 47

2.5.12 Monofilter 47

2.5.13 Nicolon 49

2.5.14 Permealiner 52

2.5.15 Petromat 53

2.5.16 Polyfelt 53

2.5.17 Poly-Filter 53

2.5.18 ProPex 56

2.5.19 Reemay 56

2.5.20 Sontara 56

2.5.21 Stabilenka 58

2.5.22 Supac 58

2.5.23 Terrafix 59

2.5.24 Terram and Filtram 60

2.5.25 Typar 63

2.5.26 Tyvek 63

2.5.27 Summary of Fabric Properties 68

2.6 References 68

3 Fabric Use in Separation of Materials *71*

3.1 Zoned Earth Dams 72

3.2 Railroad Ballast/Subgrade Separation 80

3.3 References 83

4 Fabric Use as Reinforcement *84*

4.1 Fabrics in Road Construction 87

4.2 Fabrics in Slope Stability Problems 98

Conversion Relationships

Different unit systems are used in different countries. In the United States the English system is still used although the geotechnical community has long used the metric system for most material properties. On the other hand, the contracting and textile communities continue to use the English system. Current efforts are toward implementation of the SI system (Système International d'Unités), which is based on fundamental metric units. The following relationships should allow the reader to convert from units of one system to the other. These relationships use the SI units of m/s^2 for acceleration; m/s or cm/s for velocity; mm, cm, m, or km for length; cm^2 or m^2 for area; cm^3 or m^3 for volume; liter for capacity; N (newton) for force; Pa (pascal) for pressure or stress; and degrees centigrade for temperature.

Length

1 mm = 0.0394 in.
1 cm = 10 mm = 0.394 in.
1 m = 100 cm = 39.4 in. = 3.28 ft
1 km = 1000 m = 3280 ft = 0.621 miles

1 in = 2.54 cm
1 ft = 0.305 m
1 yd = 0.914 m
1 mile = 1.609 km

Area

1 cm^2 = 0.155 in^2
1 m^2 = 10.8 ft^2 = 1.20 yd^2
1 ha = 2.47 acres

1 $in.^2$ = 6.45 cm^2
1 ft^2 = 0.0929 cm^2
1 yd^2 = 0.835 m^2
1 acre = 0.405 ha

Volume

$1\ \text{cm}^3 = 0.0610\ \text{in}^3$
$1\ \text{m}^3 = 35.3\ \text{ft}^3 = 1.31\ \text{yd}^3$

$1\ \text{in.}^3 = 16.4\ \text{cm}^3$
$1\ \text{ft}^3 = 0.0283\ \text{m}^3$
$1\ \text{yd}^3 = 0.764\ \text{m}^3$

Capacity

$1\ \text{liter} = 1000\ \text{cm}^3$
$1\ \text{liter} = 61.0\ \text{in}^3$
1 liter = 0.264 U.S. gallon
$1\ \text{U.S. gallon} = 3785\ \text{cm}^3$
$1\ \text{U.S. gallon} = 231\ \text{in}^3$
1 U.S. gallon = 3.78 liter

$1\ \text{cm}^3 = 0.001\ \text{liter} = 2.64 \times 10^{-4}\ \text{U.S. gallon}$
$1\ \text{ft}^3 = 7.48\ \text{U.S. gallon} = 28.3\ \text{liter}$

Force

$1\ \text{N} = 102.0\ \text{g} = 0.225\ \text{lb} = 1.124 \times 10^{-4}\ \text{ton}$
$1\ \text{g} = 9.81 \times 10^{-3}\ \text{N} = 2.20 \times 10^{-3}\ \text{lb} = 1.102 \times 10^{-6}\ \text{ton}$
$1\ \text{lb} = 4.45\ \text{N} = 453.6\ \text{g} = 5.00 \times 10^{-4}\ \text{ton}$
$1\ \text{ton} = 8.89 \times 10^3\ \text{N} = 9.07 \times 10^5\ \text{g} = 2000\ \text{lb}$

Stress

$1\ \text{N/m}^2 = 1\ \text{Pa}$
$= 1.02 \times 10^{-5}\ \text{kg/cm}^2 = 1.45 \times 10^{-4}\ \text{lb/in}^2 = 2.08 \times 10^{-2}\ \text{lb/ft}^2 = 1.04 \times 10^{-5}\ \text{ton/ft}$
$1\ \text{kg/cm}^2 = 9.81 \times 10^4\ \text{N/m}^2 = 14.2\ \text{lb/in}^2 = 2.05 \times 10^3\ \text{lb/ft}^2$
$= 1.02\ \text{ton/ft}^2$
$1\ \text{lb/in.}^2 = 6.89 \times 10^3\ \text{N/m}^2 = 7.03 \times 10^{-2}\ \text{kg/cm}^2 = 144\ \text{lb/ft}^2$
$= 7.2 \times 10^{-2}\ \text{ton/ft}^2$
$1\ \text{lb/ft}^2 = 4.79 \times 10\ \text{N/m}^2 = 4.88 \times 10^{-4}\ \text{kg/cm}^2$
$= 6.94 \times 10^{-3}\ \text{lb/in}^2 = 5.00 \times 10^{-4}\ \text{ton/ft}^2$
$1\ \text{ton/ft}^2 = 9.58 \times 10^4\ \text{N/m}^2 = 9.76 \times 10^{-1}\ \text{kg/cm}^2$
$= 13.9\ \text{lb/in}^2 = 2000\ \text{lb/ft}^2$

Unit Weight

$1\ \text{N/m}^3 = 1.02 \times 10^{-4}\ \text{g/cm}^3 = 6.37 \times 10^{-3}\ \text{lb/ft}^3$
$1\ \text{g/cm}^3 = 9.81 \times 10^3\ \text{N/m}^3 = 62.4\ \text{lb/ft}^3$
$1\ \text{lb/ft}^3 = 1.57 \times 10^2\ \text{N/m}^3 = 1.60 \times 10^{-2}\ \text{g/cm}^3$

Temperature

$1°C = 1°K = 1.8°F$
$1°F = 0.555°C = 0.555°K$
$0°K = -273°C = -460°F$

$$
\begin{aligned}
T_C &= (5/9)\,(T_F - 32°) \\
&= T_K - 273° \\
T_K &= T_C + 273° \\
&= (T_F + 460)/1.8 \\
T_F &= (9/5)\,T_C + 32° \\
&= 1.8\,T_K - 460°
\end{aligned}
$$

Construction and Geotechnical Engineering Using Synthetic Fabrics

1

Overview and Background of Synthetic Fibers

1.1 Synthetic Fiber Statistics

Statistical information provided by the American Textile Manufacturers Institute[1] about the textile industry is presented at the beginning of this book to give the reader an idea of the industry's scope and of the role played by geotechnical fabrics. The American textile industry consists of about 5,000 companies operating an estimated 7,200 plants in 47 states. These companies make a wide variety of products—including textiles. In recent decades the industry has been concentrating in the Southeast—notably in the Carolinas, Georgia, and Alabama—and this trend is continuing. As seen in Table 1.1, the Southeast has slightly more than 40 percent of the total operating plants. The bulk of spinning, weaving, and knitting operations of the industry is also conducted in the Southeast region.

Sales from the goods produced by the nation's textile plants run to over $35 billion a year. Textile spending for new plants and equipment in the 1960s ranged between $300 million and $800 million annually and is currently near the upper level. Expenditures will probably rise in the future.

The textile industry produces about 17 billion square yards of fabric each year. These fabrics are made from natural fibers—cotton, wool, and silk—and from synthetic fibers—nylon, rayon, polyester, acrylic, and many others.

Through the history of the American textile industry, cotton was king until the 1960s, when synthetic fibers outgrew it in use. Today the textile industry produces about 12 billion pounds of fibers annually, as seen in Table 1.2. Of these fibers, over 73 percent are synthetic, while 26 percent

TABLE 1.1 Textile Plant Location

Region	Plants (%)
Southeast	41.9
Mid-Atlantic	35.8
Northeast	11.4
Midwest	6.3
West	4.6

are cotton. About 1 percent is wool and silk. End use of the fabrics is given in Table 1.3.

Production and use of synthetic fabrics (about 73 percent of the total fabric production in 1976) are the most rapidly growing segments of the textile industry. A separate organization, the Man-Made Fiber Producers Association,[2] accumulates information on this growth, as seen in Table 1.4. Note should be made that two categories are listed, that is, cellulosic and noncellulosic fibers, and that some detail according to fiber generic name (which is explained in Section 1.3) is also listed. Of particular interest in Table 1.4 is the growth of the noncellulosic fibers, since most construction fabrics are made from these fibers; for example, nylon is used for many woven fabrics and olefin and polyester for many nonwoven fabrics.

TABLE 1.2 Textile Fiber Consumption

Fabric	1977, Estimated (lb)	1976 (lb)
Synthetic	8.8 billion	8.0 billion
Cotton	3.2 billion	3.4 billion
Wool	106.0 million	145.9 million
Silk	2.5 million	2.5 million

TABLE 1.3. End Use of Fabric

Use	Fabric (%)
Apparel	42.1
Home furnishings	30.2
Industrial fabrics	23.5
Exports	4.2

TABLE 1.4. **Synthetic Fiber Production (from Ref. 2) (in Millions of Pounds)**

Location	Year 1960	1970	1975	1976	1977
U.S.	1,630	4,959	6,621	7,566	8,200
World	7,297	18,660	25,925	27,269	29,200

U.S. Synthetic Fiber Production by Fiber[a]
(in Millions of Pounds)

Fiber	Year	Yarn and Monofilaments	Staple and Tow	Total
Cellulosic Fibers	1960			
Rayon		426.3	314.0	740.3
Acetate[b]		228.2	60.0	288.2
Noncellulosic Fibers				
Nylon[c]		384.8	26.8	411.6
Acrylic[d]		0	135.7	135.7
Olefin[e]		13.7	0	13.7
Polyester		38.9	2.1	41.0
Total				1,630.5
Cellulosic Fibers	1970			
Rayon		267.6	607.4	875.0
Acetate[b]		463.2	35.0	498.2
Noncellulosic Fibers				
Nylon[c]		1,137.9	216.8	1,354.7
Acrylic[d]		0	491.9	491.9
Olefin[e]		200.4	58.8	259.2
Polyester		455.1	1,022.3	1,477.4
Other[f]			3.0	3.0
Total				4,959.4
Cellulosic Fibers	1975			
Rayon		64.8	370.9	435.7
Acetate[b]		286.9	12.0	298.9
Noncellulosic Fibers				
Nylon[c]		1,295.4	561.9	1,857.3
Acrylic[d]		0	524.6	524.6
Olefin[e]		442.9	53.0	495.9
Polyester		1,458.9	1,536.2	2,995.1
Other[f]		11.7	2.1	13.8
Total				6,621.3

TABLE 1.4. (*Continued*)

Fiber	Year	Yarn and Monofilaments	Staple and Tow	Total
Cellulosic Fibers	1976			
Rayon		74.8	475.4	550.2
Acetate[b]		286.9	11.0	397.9
Noncellulosic Fibers				
Nylon		1,372.5	702.7	2,075.2
Acrylic[d]		0	621.0	621.0
Olefin[e]		509.3	56.5	565.8
Polyester		1,404.7	1,935.6	3,340.3
Other[f]		12.0	3.3	15.3
Total				7,565.7

[a] Figures in this table exclude textile glass fibers.
[b] Acetate includes diacetate and triacetate. Does not include poundages produced for cigarette filtration purposes.
[c] Nylon includes aramid.
[d] Acrylic includes modacrylic.
[e] Olefin includes silt and split fiber.
[f] Other includes anidex, spandex, saran, TFE-Fluorocarbon, and vinyon.

1.2 Synthetic Fiber Production

1.2.1 The Fiber Base

Most synthetic fibers are formed by forcing a syrupy substance (about the consistency of honey) through the tiny holes of a device called a spinneret.

In their original state, the fiber-forming substances exist as solids and therefore must first be converted into a liquid state for extrusion. This is achieved by dissolving them in a solvent or by melting them with heat. If they cannot be dissolved or melted directly, they must be chemically converted into soluble derivatives.

The basic substance for the three cellulosic fibers (acetate, rayon, and triacetate) is cellulose, which comes from purified wood pulp. It can be dissolved for extrusion into fibers. The substances used in the production of noncellulosic fibers generally are melted or chemically converted into a liquid state.

Unlike natural fibers, the synthetics can be extruded in different thicknesses. This is called denier. It is the industry's word for measuring the size of a continuous monofilament, a multifilament yarn, or cut staple fiber. Denier is defined as the weight in grams of 9,000 meters of fiber.

Fifteen denier monofilament is commonly used in hosiery to achieve ultimate sheerness, whereas fibers of 840 denier are used in tires for trucks, automobiles, planes, and other vehicles, giving greater strength. A term related to denier is tex. Tex is defined as the weight in grams of 1,000 meters of fiber and, as such, it is equal to the denier divided by nine.

1.2.2 The Spinneret and Related Processing

The spinneret, which is used in the production of synthetic fibers, is similar in principle to the shower head in a bathroom—liquid is forced through the holes. A spinneret can have from one to literally thousands of tiny holes and is generally made from corrosion-resistant metals.

The filaments emerging from the holes in the spinneret are then hardened or solidified. The process of extrusion and hardening is called spinning, not to be confused with the textile operation of the same name. There are three methods of spinning synthetic fibers; wet, dry, and melt spinning. Some fibers may be produced by more than one method.

When the filaments, as they emerge from the spinneret, pass directly into a chemical bath where they are solidified or regenerated, the process is called wet spinning (because of the bath). Acrylic and rayon are produced by this process.

When filaments coming from the spinneret are solidified by being dried in warm air, it is called dry spinning. This process is used in the production of acetate, acrylic, modacrylic, spandex, triacetate, and vinyon.

When the fiber-forming substance is melted for extrusion and hardened by cooling, the process is called melt spinning. Nylon, olefin, polyester, aramid, and glass are produced by the melt spinning process.

1.2.3 Subsequent Operations

While the fibers are hardening, or after they have been hardened, they are stretched. This reduces the fiber diameter, or denier, and causes the molecules in the fiber to arrange themselves into a more orderly pattern. In a given fiber type, the strength increases and the fiber's ability to stretch without breaking decreases as the pattern of the molecular arrangement becomes better oriented. A wide range of strength-to-stretch combinations may be produced in this way.

Staple fibers are produced by first extruding many continuous filaments of specific denier from the spinneret in a large ropelike bundle called tow (see Figure 1.1). A tow may often contain as many as 200,000 continuous filaments. These big bundles of fibers are crimped and then

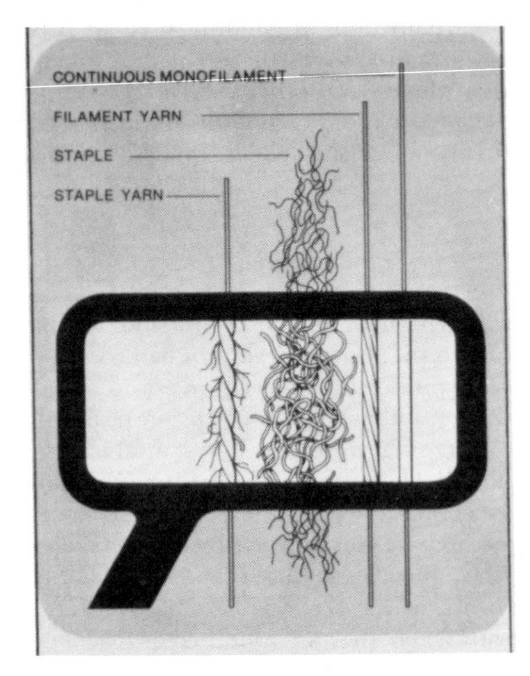

FIGURE 1.1. Enlarged view of yarns (from Ref. 2).

mechanically cut into the desired short staple lengths, usually one to four inches.

Synthetic fiber yarns can be textured by special machinery, giving them bulk, stretch, and greater comfort qualities. The special texturing machines twist and manipulate the continuous filament yarns in such a way that the filaments no longer lie exactly parallel to one another. The increased space between the filaments allows the development of the special qualities mentioned above.

Combination yarns can be formed by combining spun staple and filament yarns in numerous ways. These yarns give added strength. They are sometimes used as additional reinforcement wherever needed.

The short fibers, or staple, may be twisted or spun, just as short lengths of natural fibers are spun. Staples of various lengths and denier are designed for use in various systems of spinning. The principal spinning systems used are: (1) the cotton system, (2) the wool system, or (3) the worsted system. Some staple (usually crimped) is used without spinning—as filling in pillows, mattresses, sleeping bags, and comforters. This is called fiberfill.

Yarns spun from staple are more irregular than filament yarns. The short ends of fibers, projecting from the yarn surface, produce a fuzzy effect. Spun yarns are also more bulky than filament yarns of the same weight. They are, therefore, more often used for porous, warm fabrics and for the creation of nonsmooth surfaces for fabrics.

Synthetic fibers can be blended with other fibers, either natural or synthetic. When two or more types of staple fibers are blended, they bring together the properties common to each into a single yarn.

In some cases, a wide sheet of film is extruded and subsequently slit lengthwise into narrow continuous strips which, depending upon width, could be correctly described as monofilaments. These slit filaments are sometimes combined and used as multifilament yarns.

Synthetic fibers can also be extruded from the spinneret in different shapes (round, trilobal, pentagonal, octagonal, and others), whereas natural fibers are available only in the form that nature provides. Trilobal fibers reflect more light and give an attractive sparkle to textiles. Pentagonal fibers, when used in carpets, show less soil and dirt. Octagonal fibers offer glitter-free effects.

With synthetic fibers, the concept of blending or combining different materials can actually be taken all the way back to the extrusion process. Two different polymers can be extruded side by side in a single fiber coming from the spinneret to create a bicomponent. One of the polymers will have greater heat and/or moisture sensitivity than the other and will spiral during the finishing process, thus creating a fiber with greater bulk and comfort.

Two different polymers can be homogeneously mixed together during or prior to extrusion from the spinneret. This combines the characteristics from the two materials into a single fiber, called a biconstituent.

It also is possible to add certain additives to the polymers or the solution before it is extruded, thereby giving the finished fiber special characteristics, such as antistatic or flame retardant properties. For example, the process of adding color to the polymer prior to extrusion is called solution dyeing; it gives a high degree of colorfastness. Synthetic fibers also can be cross dyed. This is a fascinating technique in which cloth or carpet is woven, knitted, or tufted from different types of the same generic fibers, then put through a dye bath. The different type fibers react to the different colors in the dye bath, creating a special color. This enables a mill to offer its customers many different color combinations of a desired pattern.

1.2.4 Forms of Synthetic Fibers

From these processes many different fiber forms result. For example, monofilament yarn is a single filament (fine thread) of continuous length, as seen in Figure 1.1. Multifilament yarn is two or more continuous monofilaments assembled or held together by twist or otherwise. Tow is a large bundle of continuous monofilaments assembled with twist, while staple is discontinuous lengths of fibers that have been cut or broken into desired lengths from large bundles of continuous monofilaments, or tow (see Figure 1.1). An excellent treatment of the entire field is given by Kaswell in the *Handbook of Industrial Textiles.*[3]

1.3 Generic Names and Variants (Trademarks)

The *generic* name of a synthetic fiber is similar to a family name. Each generic fiber has specific basic characteristics that perform needed functions in textile products. The Federal Trade Commission's definitions (see Appendix 10.1) describe the broad, general chemical composition of a generic fiber.

Currently, there are twenty-one generic names for synthetic fibers. Table 1.5 lists the current ones with the dates of their first commercial United States production.

Anidex, axlon, lastrile, novoloid, nytril, and vinal are also generic names but are not currently produced in the United States. The definitions of these variants are also given in Appendix 10.1.[2]

It is possible to work within the basic generic composition and modify it, both chemically and physically, in order to produce a wide variety of different fibers, called *variants.* These are members of a generic fiber family, each with different characteristics, yet all conforming to a basic Federal Trade Commission generic definition. Fiber variants are generally developed for a special purpose. They may be engineered to offer toughness, heat resistance, or to reduce degradation. They may offer soil

TABLE 1.5. Various Generic Names

Date	Generic Name	Date	Generic Name
1910	Rayon	1949	Modacrylic
1924	Acetate	1949	Olefin
1930	Rubber	1950	Acrylic
1936	Glass	1953	Polyester
1939	Nylon	1954	Triacetate
1939	Vinyon	1959	Spandex
1941	Saran	1967	Aramid
1946	Metallic		

TABLE 1.6. Some Physical Properties of Synthetic Fibers (Standard Laboratory Conditions for Fiber Tests: 70°F and 65% Relative Humidity)[a]

Fiber	Breaking Tenacity[b] (g/denier)		Specific Gravity[c]	Standard Moisture Regain (%)[d]	Effects of Heat
	Standard	Wet			
Acetate (filament and staple)	1.2–1.5	0.8–1.2	1.32	6.0	Sticks at 350°F (177°C) to 375°F (191°C); softens at 400°F (205°C) to 445°F (230°C); melts at 500°F (260°C); burns relatively slowly
Acrylic (filament and staple)	2.0–3.5	1.8–3.3	1.14–1.19	1.3–2.5	Sticks at 450°F (232°C) to 497°F (258°C), depending on type
Aramid					
Regular tenacity filament	4.8	4.8	1.38	5	Decomposes above 800°F (427°C)
High tenacity filament	22	22	1.44	2.7–7	Decomposes above 900°F (482°C)
Staple	3–4.5	3–4.5	1.38	5	Decomposes above 800°F (427°C)
Modacrylic (filament and staple)	2.0–3.5	2.0–3.5	1.30–1.37	0.4–4.0	Will not support combustion; shrinks at 250°F (121°C); stiffens at temperatures over 300°F (149°C)
Nylon					
Nylon 66 (regular tenacity filament)	3.0–6.0	2.6–5.4	1.14	4.0–4.5	Sticks at 445°F (229°C); melts at about 500°F (260°C)
Nylon 66 (high tenacity filament)	6.0–9.5	5.0–8.0	1.14	4.0–4.5	Same as above

TABLE 1.6. *(Continued)*

Fiber	Breaking Tenacity[b] (g/denier)		Specific Gravity[c]	Standard Moisture Regain (%)[d]	Effects of Heat
	Standard	Wet			
Nylon 66 (staple)	3.5–7.2	3.2–6.5	1.14	4.0–4.5	Same as above
Nylon 6 (filament)	6.0–9.5	5.0–8.0	1.14	4.5	Melts at 414°F (212°C) to 428°F (220°C)
Nylon 6 (staple)	2.5	2.0	1.14	4.5	Melts at 414°F (212°C) to 428°F (220°C)
Olefin (polypropylene) (filament and staple)	4.8–7.0	4.8–7.0	0.91	—	Melts at 325°F (163°C) to 335°F (168°C)
Polyester					
Regular tenacity filament	4.0–5.0	4.0–5.0	1.22 or 1.38[e]	0.4 or 0.8[e]	Melts at 480°F (249°C) to 550°F (288°C)
High tenacity filament	6.3–9.5	6.2–9.4	1.22 or 1.38[e]	0.4 or 0.8[e]	Melts at 480°F (249°C) to 550°F (288°C)
Regular tenacity staple	2.5–5.0	2.5–5.0	1.22 or 1.38[e]	0.4 or 0.8[e]	Melts at 480°F (249°C) to 550°F (288°C)
High tenacity staple	5.0–6.5	5.0–6.4	1.22 or 1.38[e]	0.4 or 0.8[e]	Melts at 480°F (249°C) to 550°F (288°C)
Rayon (filament and staple)					
Regular tenacity	0.73–2.6	0.7–1.8	1.50–1.53	13	Does not melt. Decomposes at 350°F (177°C) to 464°F (240°C)
Medium tenacity	2.4–3.2	1.2–1.9	1.50–1.53	13	Burns readily
High tenacity	3.0–6.0	1.9–4.6	1.50–1.53	13	
High wet modulus	2.5–5.5	1.8–4.0	1.50–1.53	13	

Spandex (filament)	0.6–0.9	0.6–0.9	1.20–1.21	0.75–1.3	Degrades slowly at temperatures over 300°F (149°C); melts at 446°F (230°C) to 519°F (270°C)
Saran (filament)	up to 1.5	up to 1.5	1.70	—	Softens at 240°F (116°C) to 280°F (138°C); self-extinguishing
Triacetate (filament and staple)	1.2–1.4	0.8–1.0	1.3	3.2	Before heat treatment, sticks at 350°F (177°C) to 375°F (191°C); after treatment, sticks above 464°F (240°C); melts at 575°F (302°C)
Vinyon (staple)	0.7–1.0	0.7–1.0	1.33–1.35	up to 0.5	Becomes tacky and shrinks at 150°F (66°C); softens at 170°F (77°C); melts at 260°F (127°C); will not support combustion

[a] Data given in ranges may fluctuate according to introduction of fiber modifications or additions and deletions of fiber types.
[b] Breaking tenacity: the stress at which a fiber breaks, expressed in terms of grams per denier.
[c] Specific gravity: the ratio of the weight of a given volume of fiber to an equal volume of water.
[d] Standard moisture regain: the moisture regain of a fiber (expressed as a percentage of the moisture-free weight) at 70°F and 65% relative humidity.
[e] Depending on type.

release or antistatic properties. Variants also may be created for better blending with other fibers to impart different properties to a textile material. There are endless possibilities for modifying a basic generic structure and producing new fibers with specific properties. The ability to engineer fibers with special built-in qualities is one of the truly unique aspects of the synthetic fiber industry. This has resulted in many improved textile products for the consumer.

When a fiber manufacturer develops a new variant, it is usually given a name. This is called a *trademark* and is owned and promoted by the fiber manufacturer who produced it. Some trademarks are better known than others, just as some generic names are more familiar to the public. It is important to clearly understand the difference between a trademark and a generic name. The most common trademarks for current generic classifications of member companies of the Man-Made Fibers Producers Association are listed in Appendix 10.2.[2] There are many more in addition to those listed.

1.4 Some Physical Properties of Synthetic Fibers

Since the construction use of fabrics in this book depends heavily on the physical and mechanical properties of the materials involved, it is necessary to briefly consider the physical and mechanical properties of the fibers from which the fabrics are made (see Table 1.6). Particular concern should be directed to nylon, olefin (polypropylene), and polyester, which are used to make the majority of construction fabrics currently in use.

1.5 References

1. American Textile Manufacturers Institute, Inc., 1501 Johnston Building, Charlotte, N.C. 28281.
2. Man-Made Fiber Producers Assoc., Inc., Education Department, 1150 Seventeenth Street, N.W., Washington, D.C. 20036.
3. E. R. Kaswell, *Handbook of Industrial Textiles*, Wellington Sears, New York, 1963, 757 pp.

2

Construction Fabrics

2.1 Elements of Fabric Construction

Modern industrial textiles are created by textile engineers using precise scientific methods to produce fabrics that will properly perform specific tasks. In each case, it is first necessary to project a clear picture of the final application itself, bringing into sharp focus all the requirements to be met in actual use. With the objective thus clearly established, every step in engineering the fabric is aimed at achieving superior end use performance.[1]

To accomplish this purpose, the textile engineer works with five basic elements of fabric construction: (1) fibers, (2) yarn, (3) weave, (4) count, and (5) finish. All of these elements must be accurately evaluated and controlled, individually and in combination, for successful construction of a fabric. The resulting *woven* fabrics are discussed in Section 2.1.1 and shown in Figure 2.1.

It should be noted at the outset, however, that many construction fabrics are *nonwoven*. Thus the elements of yarn and weave are completely lacking in these bonded fabrics. Neither felt nor paper, these fabrics are composed of textile fibers bonded together by resins, other bonding agents, or mechanical methods to form a smooth, uniform mat. They are discussed in Section 2.1.2 and shown in Figure 2.1. Also shown in Figure 2.1 is an example of *knit* fabric, which, although not used in construction fabrics, gives perspective to the overall range of fabric types.

2.1.1 Woven Fabrics

Obviously the choice of proper fiber is all important in the formation of any fabric. For this reason it is singled out in Chapter 1 as a separate topic. The fibers are then used to form yarn, which is the second element in the formation of a fabric. It, too, is discussed briefly in Chapter 1.

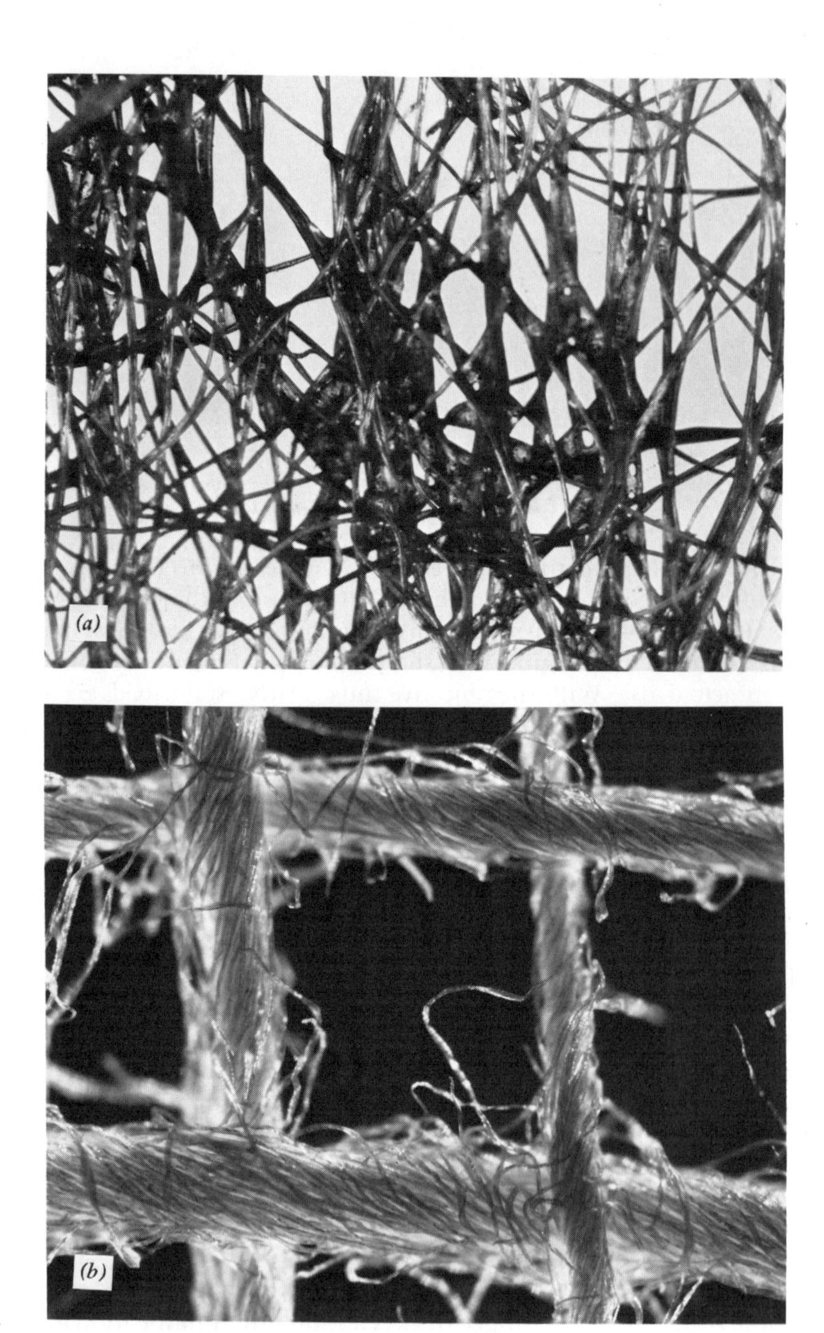

FIGURE 2.1. Photomicrographs of (a) dryformed nonwoven fabric, (b) fabric woven of yarns spun from staple fibers, and (c) knit fabric made of filament yarns (from Ref. 2).

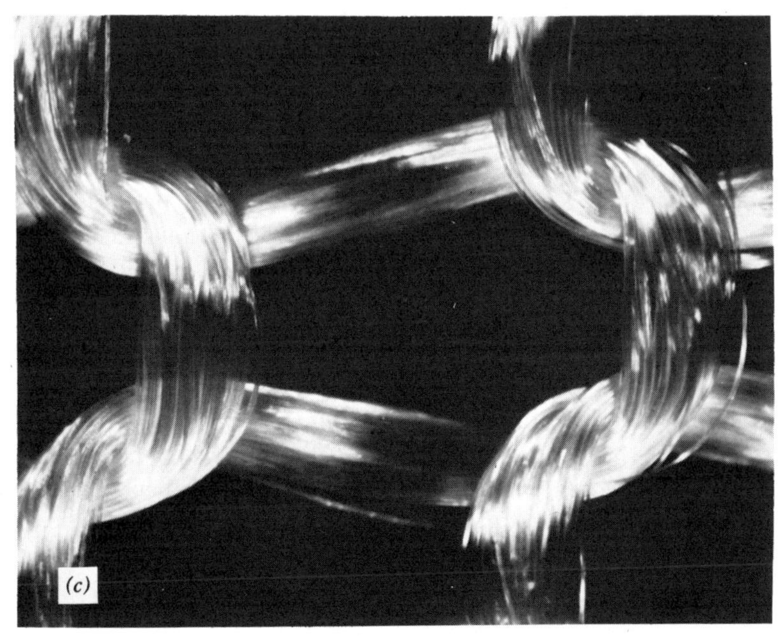

FIGURE 2.1. *Continued.*

The actual formation of a woven fabric begins with the weave, that is, the third element in the process. Weave is the system of interlacing lengthwise yarns ("ends") with crosswise filling yarns ("picks"). While there are many varieties of weaves, all are developed from the three basic weaves: plain, twill, and satin.

Plain weave is simply "one up and one down." In a tight construction the plain weave can provide more interlacings per square inch than any other weave, an important factor when impermeability and covering qualities are required. Yet it can be opened up to practically any desired degree, thus producing various degrees of permeability.

Among heavy plain weave fabrics are cotton ducks: "flat" ducks, so-called because two warp ends are woven flat together as one, side by side but not plied; army duck, "numbered" ducks, hose and belting ducks, all made of plied yarns; chafer, filter, and other special ducks. Sheeting and fine-combed cottons such as lawns, airplane and balloon cloth, and typewriter ribbon fabric are other examples of plain weave fabrics.

Twill weave is easy to recognize by the sharp diagonal "twill" line caused by each end crossing two or more picks with the interlacing advancing one pick with each warp end. The twill weave has fewer interlacings than the plain weave and, as a result, greater porosity. Twill weave is used in twills, drills, and jeans.

Satin weave has still fewer interlacings than the twill weave. They are widely but regularly spaced, providing a smooth surface, which is advantageous for coated materials such as upholstery. Its porosity and high cover factor enable satin weave fabrics to hold back small particles, as in dust filtration. Originally used with silk, satin weave is known as "sateen" when used with cotton.

Among the many variations that are derived from these three basic weaves, the following are useful in industrial applications. Leno weave is a plain weave variation, open and netlike, with warps interlocked to prevent slippage. It is used for laundry and dye nets and rubber reinforcement. Basket weave, with two or more yarns in warp and filling woven as one, is highly tear resistant and is used in various combinations with rubber.

Count, which is the fourth element in woven fabric construction, is the number of yarns per inch of fabric. It is a vital consideration in the construction of woven fabrics because it affects absorbency, adhesion, permeability, strength, weight, bulk, flexibility, and other fabric characteristics. The "gray" or unfinished fabric count, however, may be misleading because of dimensional changes due to finishing or to tension and other factors in subsequent processing.

Because good looks are not generally a requirement for industrial fabrics, many fabrics are used "in the gray"—without any finishing process. When the fabric is finished, however, it forms the fifth, and last, step in the construction of the woven fabric. Finishing steps include bleaching or dyeing, and scouring to remove oils, waxes, or sizing. Preshrinking of cotton or "heat setting" of synthetics may be necessary for dimensional stability. The effectiveness of heat setting depends on the original stabilization temperature used. If the synthetic fabric is later exposed to a significantly greater temperature, dimensional change can occur. Calendering, another finishing process, produces a thinner gauge and smoother surface. Napping is used to raise a flannel-like pile on one or both sides of the fabric. Singeing is used to remove whiskery surface fibers. Predipping is required for certain rubber applications. In addition to these, there are many other finishing processes used for industrial fabrics for specific end uses.

Since the nonwovens form an important part of construction fabrics, their manufacture is described in some amount of detail.

2.1.2 Nonwoven Fabrics

INDA, The Association of The Nonwoven Fabrics Industry, provides information about this important category of synthetics.[2] It is the group from which most construction fabrics are made. The field of nonwovens

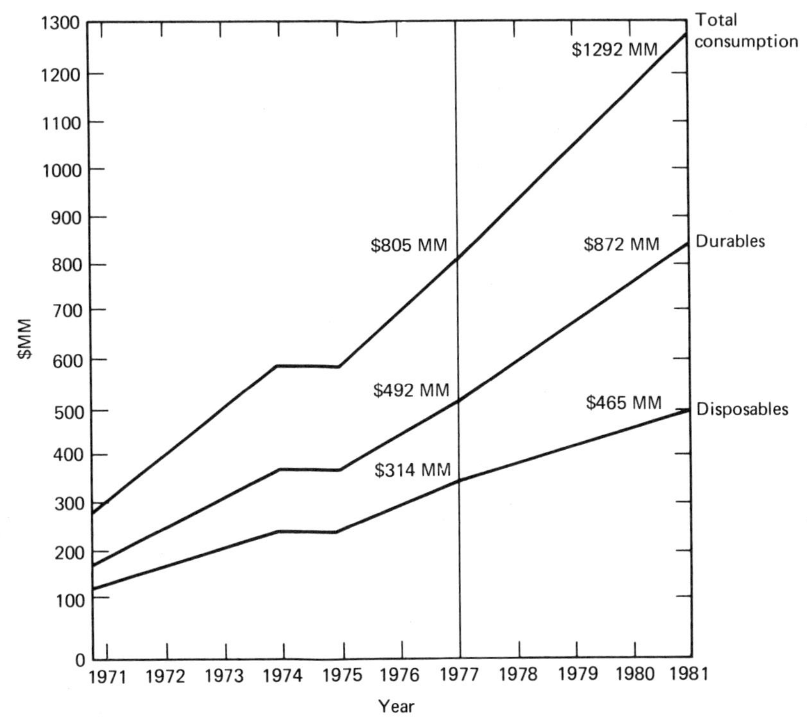

FIGURE 2.2. United States consumption of nonwoven fabrics; 1971–1981 (from Ref. 2).

represents one of the fastest-growing industries worldwide. In the United States alone, the value of nonwoven fabrics consumption has grown rapidly from less than $300 million annually at the beginning of the 1970s to more than $800 million in 1977. Estimates place 1981 nonwovens usage at $1.3 billion.

Major uses of nonwovens are disposable diapers, coated fabrics, medical and surgical supplies, carpet underlay, interlinings, industrial and household wipes, wet wipes, and civil engineering materials. Market trends are shown in Figure 2.2.

Each nonwoven manufacturing system generally includes four basic steps: (1) fiber preparation, (2) web formation, (3) web bonding, and (4) post-treatment. The processes listed in Table 2.1 indicate the various steps that may take place during each of the four basic steps.

While this maze of processes might seem overwhelming, only a few represent the majority of nonwovens used in civil engineering construction: needle punched, spun bonded, and various offshoots thereof. The following information is adapted from INDA[2] and McGown.[3]

TABLE 2.1 Elements in the Formation of Nonwoven Fabrics

Fiber preparation	Web Formation	Web Bonding	Post-Treatment
Opening	Card or garnett (unidirectional & cross laid)	Saturation	Drying
Conditioning	Air laid random	Print bonding	Curing
Blending	Wet laid random	Point bonding	Finishing
Chip preparation	Filament extrusion (spun bonding)	Foam	Printing
	Film extrusion	Wet Roll	Embossing
		Needle punching	Dyeing
		Spray	
		Solvent	
		Powder	
		Thermoplastic fiber	
		Hot calendering	
		Lamination	
		Apertured webs	
		Jet entanglement (spun lacing)	

Source: Ref. 2.

NEEDLE PUNCHED PROCESS. A needle punched fabric is produced by introducing a fibrous web—already formed by cards, garnetts, or air laying—into a machine equipped with groups of specially designed needles. While the web is trapped between a bed plate and a stripper plate the needles punch through it and reorient the fibers so that mechanical bonding is achieved among the individual fibers. Often, the batt of fibers is carried into the needle punching section of the machine on a lightweight support material or substrate. This is done to improve finished fabric strength and integrity (see Figure 2.3).

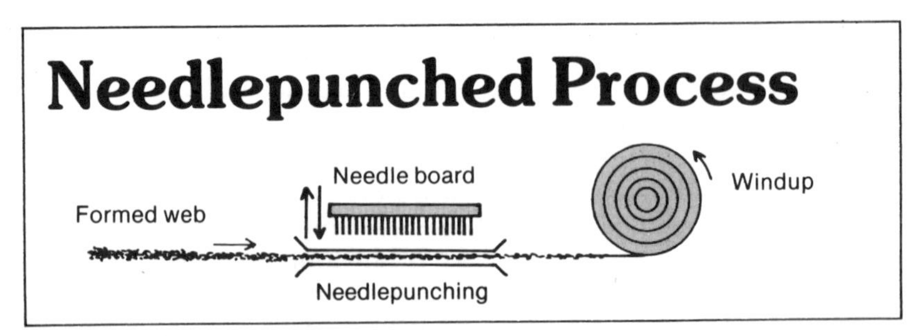

FIGURE 2.3. Schematic diagram of needle punched process (from Ref. 2).

The needle punching process is generally used to produce fabrics that have high density yet retain some bulk. Fabric weights usually range from 1.7 to 10 ounces per square yard, thicknesses from 15 to 160 mils. (Note: 1 mil = 0.001 inches is often used by the textile industry to measure fabric thickness.) Typical end uses are blankets, filter media, coated fabric backings, carpeting and carpet backings, automobile landau top substrates, apparel interlinings, road underlay, and auto trunk liners.

SPUN-BONDED PROCESS. Spun bonding is a continuous process producing a finished fabric from a polymer. A polymer, or several polymers, such as polyester, polyamide, polypropylene, polyethylene, or others, is fed into an extruder. As it flows from the extruder it is forced through a spinneret, or series of spinnerets. After cooling, the resulting continuous filaments are then laid on a moving conveyor belt to form a continuous web. In the lay-down process, the desired orientation of the fibers is achieved by various means, such as rotation of the spinneret, electrical charges, introduction of controlled air streams, or varying the speed of the conveyor belt. The fabric is then bonded by thermal, mechanical, or chemical treatment before being wound up into finished roll form (see Figure 2.4).

A wide range of fabric characteristics can be achieved by controlling the various elements in this process. High-performance low-weight fabrics are characteristic of this process because of the continuous nature of the fibers. Fabric weights usually range from 0.3 to 6 ounces per square yard. Thicknesses generally range from 3 to 25 mils. Typical end uses are coated and laminated fabrics, carpet underlay, packaging material, durable papers, sanitary napkin and diaper covers, road underlay, filtration, interlinings, disposable apparel, wall coverings, battery separators, and building materials.

MELT-BONDED PROCESS. These fabrics consist of continuous filaments or long staple fibers that are melt bonded together at filament or fiber crossover points. The resultant fabrics are rather tough and compact in appearance. Higher product strength can be achieved with this type of manufacture at lower fabric weights (60 to 399 grams per square meter) than for needle punched fabrics owing to the fiber bonding utilized in the process. The bonding operations, in fact, differ between the commercially available fabrics, depending upon the basic fiber characteristics. The bonding methods include: (1) homofil bonding, in which all filaments are composed of a single polymer type, but some of the filaments have different melting characteristics. Bonding is achieved by a high-temperature calendering operation, accomplished by passing fabric

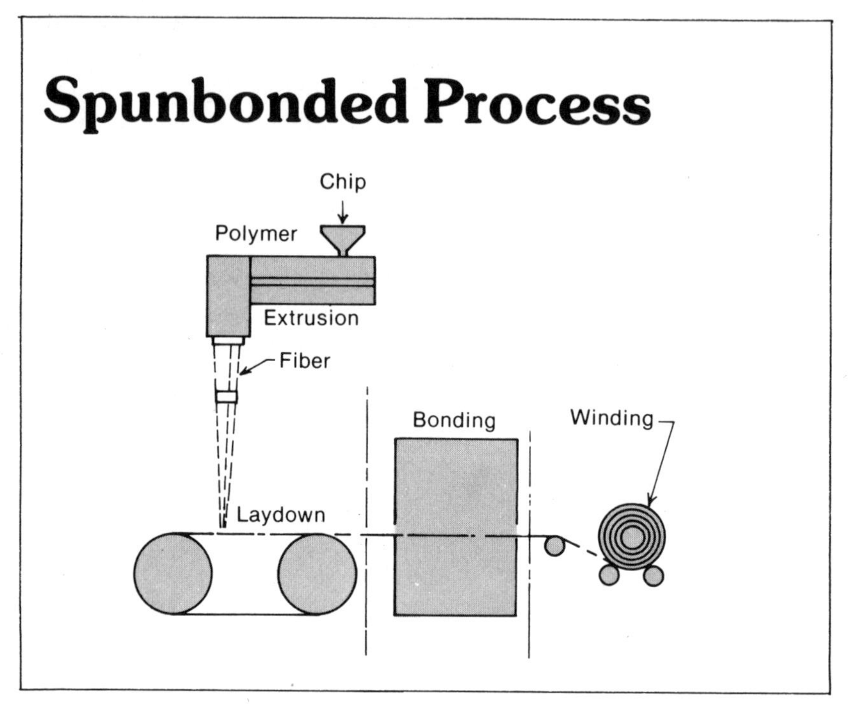

FIGURE 2.4. Schematic diagram of spun-bonded process (from Ref. 2).

between two hot rollers; and (2) heterofil bonding, in which some of the filaments comprise two types of polymers having different melting points (heterofilaments), while others have only one polymer (homofilaments). The heterofilaments have different softening characteristics from the homofilaments, and strong flexible bonds can therefore be formed at the heterofilament crossover points by controlled application of heat and pressure to fuse only the lower melting point polymer forming the sheath, while leaving the core and homofilaments unaffected.

RESIN BONDING PROCESS. Usually acrylic resin is sprayed onto, or impregnated into, a fibrous web. After curing and/or calendering, strong bonds are formed between filaments.

OTHER PROCESSES. Several other fabrics are used in civil engineering, and some exhibit such features as thread line, introduced during the manufacturing process. These are generally not directly related to the manufacturing process nor to their civil engineering performance, but are found in some quite different end uses, such as carpet backing. They can, however, have the effects of assisting strength in particular directions, and reducing strains to failure. Also, some fabrics that are now appearing are combinations of different processes, for example, a nonwoven mat needled into a woven base. The possible combinations are many, and no doubt several will be introduced into civil engineering in time.

2.2 Overview of Construction Fabric Uses

The use of natural fabrics in the construction industry probably occurred decades (or centuries) ago when some innovative individual with the idea of strengthening the material, decided to embed it within soil. The concept of embedding straw and branches in clay could even be considered to fall within this category. To trace fabric usage into modern times, however, is most difficult and can only be accurately done on the basis of industry sales on a company-to-company basis. This information not being readily available,* an alternative indicator is needed. One candidate is the number of published articles on fabrics and their uses.

Such an assessment has recently been made as part of a Federal Highway Administration grant to Bell and Hicks[5] of Oregon State University. Lewis' bibliography[6] allows for plotting of the curve shown in Figure 2.5, where the rapid increase in number of publications since 1971 is easily seen.

The year 1977 was particularly significant because of a conference in Paris in November 1977 organized by LCPC and ENPC, the International Conference on the Use of Fabrics in Geotechnics (Soils). The proceedings[7] of this conference produced 66 papers, which is 22 percent of the total listed in the Oregon State survey.[6] The Conference organized sessions along the following topic areas:

- Fabrics at the subsoil-construction material interface (roads, railways, etc.).
- Fabrics at the subsoil-construction material interface (earth fills).
- Multilayer soil-fabric systems.
- Mechanisms of soil-fabric interaction.
- Functioning of fabrics, including filters.
- Drainage.
- Consolidation.
- Tests and specifications.

*Although specific data are not available, one source cites that 6,000,000 square yards of fabric were used in construction in 1976 and estimates that more than 15,000,000 square yards were used in 1977.[4] An estimated 75,000,000 square yards were used in 1978.

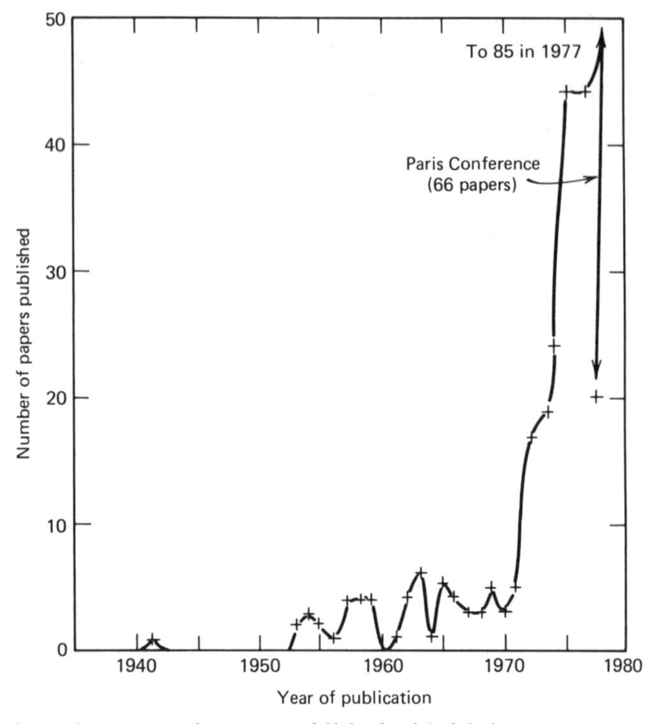

FIGURE 2.5. Summary of papers published with fabric as the main topic area.

Considerable thought was given to a similar type of structure for this book, but we felt that construction-oriented people would feel more at ease with the following structure, which is the one that is used herein:

- Separation.
- Reinforcement.
- Drainage.
- Erosion control.
- Forms.
- Impermeable fabrics.

These six specific topics are treated completely in Chapters 3 through 8, respectively, and only a brief overview is described at this time.

2.2.1 Separation

Fabrics are used in the context of separation by keeping two dissimilar materials apart. There are many construction applications where this is important. The role for fabrics, of course, is to do this task more economically and/or better than other current methods. Typical application areas are as follows:

- Separation of zoned sections of dissimilar materials within an embankment, earth dam, or rock-filled dam.
- Separation of stone base from subgrade beneath airfield and highway pavements, parking lots, secondary roads, sidewalks, and so on, that is, prevention of intrusion.
- Separation of railroad ballast from soil subgrade or of railroad ties from ballast, that is, prevention of intrusion.
- Separation of stone or other material that is placed on a temporary basis to be subsequently removed, for example, surcharge loads for soft soils or downstream berms for unstable slopes.
- Separation of frost-susceptible soils into two distinct layers, thereby breaking continuity of the capillary flow zone.

2.2.2 Reinforcement

Problems involving subgrade reinforcement appear to be particularly well suited for fabric utilization. Currently, most fabrics appear to be used in this type of application. The concept is theoretically sound since the fabric decreases the level of stress in the foundation soil due to horizontal shear stresses mobilized by the vertical loads. This in turn places the fabric in tension (similar in action to a prestressing tendon in reinforced concrete), which spreads the load over a large area and thereby decreases its intensity. That is, the unit vertical stress is decreased. A decrease in stress means less likelihood of a failure and/or less settlement. Examples of fabrics being used in this regard are as follows:

- Building of temporary roads over marshes, swamps, peat soils, or compressible fine-grained soils.
- Building of parking lots or storage handling sites over similar poor soil conditions.
- Building of facilities of almost any nature over permafrost, muskeg, and other soils in cold weather regions.
- All of the separation problem areas listed in Section 2.2.1 where the *in situ* soil needs reinforcement. In these cases the fabric is acting as both a separator and a reinforcement material.
- Construction of fabric walls and reinforcing selected zones.
- Reduction of the need for removing existing soil (of relatively poor characteristics) in marginal situations.

- Increase in the stability of embankments and dams.
- Containment of soils that would spread laterally if left unreinforced.
- Reduction of crack reflection when using a new bituminous concrete overlay over a cracked, broken, or otherwise nonsuitable existing pavement.

2.2.3 Drainage

This is perhaps the second largest area of current fabric utilization because the controlled permeability of the fabric can be put to good and economical use in many drainage situations. Major areas of application for fabrics in drainage situations are as follows:

- Prevention of migration of soil fines in a crushed stone or pipe underdrain system. This has the major effect of eliminating the need for an inverted filter consisting of various sized gravel and sand layers.
- Prevention of penetration and loss of coarse material of high permeability into the adjacent soil. When combined with the previous comment it is seen that the fabric is acting in two ways.
- Elimination of the need for all or some of the sand blanket layer used in surcharging soils with sand drains by providing an exit for the water. This might even have a syphoning effect on the system.
- Elimination of the need for all or some of a graded filter in earth dam construction in chimney drain and drainage gallery areas.
- Providing an interceptor system to block the flow path of water and guide it in another direction, for example, from horizontal flow into an underdrain system.
- Providing drainage behind temporary and permanent retaining walls.

2.2.4 Erosion Control

The use of fabrics in erosion control has seen numerous applications. In many cases the fabric acts as both a separator and as a drainage layer, but when its primary function is to aid in erosion control it is placed in this separate category. Typical situations are the following:

- As shore and coastal beach protection, where the fabric acts as a mechanism to hold the soil in place while allowing for germination of vegetation and weed growth.

- As a boundary material beneath a stone layer, rip-rap, or gabions in protecting slopes adjacent to flowing water or in tidal areas.
- As protection against erosion at water and sewer outfalls.
- As erosion control mattresses to protect slopes adjacent to flowing water or in tidal areas.
- As an artificial seaweed to allow for buildup of natural sediments and plant growth.
- As a silt fencing to block migration of soil from fines being carried by water or by wind.

2.2.5 Forms

Fabric materials can act as forms to be filled with other materials and can thereby conform to the shape and topography of any surface on which they are constructed. The controllable permeability will easily allow for the escape of air or water but will contain the injected permanent material. As such it has seen many uses in the following application areas:

- As a form for concrete in repair of piles, columns, beams, and so on.
- As a form for grout of various types and consistencies for use in erosion control, wall sections, repairs to existing structures, and so on.
- As a form for soil to construct retaining walls, shore protection systems, and so on.
- As a means to shore up mines, caverns, and so on, where access from below ground is difficult or impossible.

2.2.6 Impermeable Fabrics

Once a fabric is rendered impermeable as, for example, by impregnation and/or coating with vinyl, neoprene, elastomers, polyurethane, and so forth, an entirely new set of construction uses of fabrics arises. The fabric, being nonporous, can now be lifted by air to form an air-supported structure capable of wide variations of size and shape. It can also be water filled, taking the shape of the fabric under hydrostatic pressure to form various enclosures or barriers such as inflatable flood gates in tidal zones. Lastly, if the coating has structural stiffness, it, in itself, becomes a structural element. Its use is now unlimited for enclosures, roofs, partitions, and many other related uses. Indeed, the area of impermeable fabrics is gigantic and can only be treated superficially in this book.

2.3 Fabric Properties of Importance in Construction Use

The selection of a fabric for a particular construction application must necessarily depend upon adequate and suitable fabric properties and characteristics. If these properties and characteristics are excessive an uneconomical selection is made. If these properties and characteristics are inadequate a failure will result. The safest way out of this dilemma is through comparison with similar field projects. This could be termed a case history or observational method. Unfortunately, too few case histories are available in the relatively new field of fabrics in construction for such an approach. Thus one must resort to an evaluation of fabric properties and characteristics on the basis of laboratory tests. While such a comparison between laboratory tests and field performance is inherently dissatisfying (due primarily to scale effects, boundary conditions, inadequate testing time, etc.), there does not appear to be a viable alternative.

On the more positive side, the literature on fibers and fabrics is abundant (albeit not geotechnical testing oriented or even construction fabric oriented), and laboratory testing is well established. The American Society for Testing and Materials (ASTM) currently devotes two entire volumes to textile material testing and still another volume to testing of plastics. Many of these tests cannot be used directly, but they do provide for an excellent base for extensions and adaptations to suit construction use needs. In 1977 Committee D-13 of the ASTM (on textiles) created a Subcommittee (D13.61) to evaluate, modify and/or suggest test methods for geotechnical fabrics. From this task group, three ad hoc committees were formed to look at the following areas vis-à-vis current testing methods:

- Mechanical properties.
- Hydraulic properties.
- Endurance and miscellaneous properties.

In this section we use a similar grouping (adding only physical properties) to summarize the existing and future possible tests for construction fabrics. Where use of copyrighted ASTM information is cited it is reprinted by permission of the American Society for Testing and Materials, 1916 Race St., Philadelphia, Pa., 19103.

2.3.1 Tests for Physical Properties

WEIGHT (MASS). The current ASTM test for this property is designated D 1910. Fabric weight is expressed as mass per unit area in ounces per square yard (or grams per square meter), ounces per linear yard (or grams per linear meter) or, inversely as linear yards per pound (or meters per kilogram). In the latter two designations fabric width must also be stated. Weight determination should be made to the nearest 0.01% of the specimen weight, and the length and width should be measured under zero tension.

THICKNESS. The thickness of a fabric is the distance between the upper and lower surface of the material, measured under a specified pressure. ASTM method D 1777 stipulates that the thickness is to be measured to an accuracy of at least 0.001 inch (or 0.02 millimeter) under the following pressures. It should be noted that construction fabrics are not included in Table 2.2; thus the specification needs clarification in this regard. It becomes particularly important when thickness is required in the calculation of another fabric property; for example, determining the coefficient of permeability requires the thickness of the specimen through which flow is occurring.

TABLE 2.2. ASTM Guide for Pressures to Measure Fabric Thickness

Type of Material	Example	Pressure Range
Soft	Blankets, fleeces, knits, lofty nonwovens, woolens	0.005–0.50 psi (0.35–35 g/cm^2)
Moderate	Worsted, sheeting, carpets	0.02–2.0 psi (1.4–144 g/cm^2)
Firm	Ducks, asbestos fabrics, felts	0.1–10 psi (7–700 g/cm^2)

COMPRESSIBILITY. The thickness response of a fabric to varying pressure (rather than a specific pressure) can be measured in accordance with ASTM method D 1777, as just described. In a manner similar to geotechnical testing, the slope of the resulting curve is a compressibility modulus (or coefficient), the latter being an informative physical/mechanical property.

2.3.2 Tests for Mechanical Properties

STRIP TENSILE TEST. This is a strength test described in ASTM methods D 1682 and D 751 in which the specimen is gripped in clamps for its full width and tested to failure. A 2-inch-wide strip of fabric is often used. The testing machine loading can be applied in one of three ways: constant rate of transverse, constant rate of load, or constant rate of extension. The test suffers from a Poisson's ratio effect wherein transverse dimensions are decreased as load is applied. In some fabrics the width decreases to less than 50 percent of its original width.

GRAB TENSILE TEST. This commonly used textile strength test, ASTM methods D 1682 and D 751, is one in which only a part of the width of the specimen is gripped between clamps and the specimen tested to failure. Typically, a 4-inch-wide specimen is centrally gripped in 1-inch-wide jaws. Testing can be accomplished in a constant-rate-of-transverse, constant-rate-of-load, or constant-rate-of-extension testing machine. The nonloaded portion of the strip provides for some amount of transverse stiffness.

PLANE STRAIN TENSILE TEST. Since geotechnical fabrics are customarily restrained when placed in the field, a laboratory tensile test that maintains the transverse dimension seems attractive. Sissons[8] has devised such a test in which restraint at right angles to the direction of loading is provided by means of lightweight wooden brackets in which steel pins have been set. This lateral resistance to transverse shortening results in an increased modulus and a decreased extensibility at failure. The test has deficiencies, such as high stress concentration at points where the pins puncture the fabrics, but it attempts direct field simulation, and seems to be an improvement over the more traditional tensile tests.

BIAXIAL TENSILE TEST. The logical extension of the above tests is into biaxial tension, where the specimen is usually cut into the form of a cross and is stressed in perpendicular directions.[1] In woven fabrics under biaxial test the load rates must be kept constant to prevent skewing of the fabric, but this may not be the case for nonwoven fabrics. The major use of the test seems to be in understanding the deformation properties of the fabric before failure. This is an important determination for such fabric behavior is required in computer modeling, for example, using finite element methods to design or analyze soil/fabric systems.

ELONGATION TESTS. With either strip, grab, plane strain, or biaxial tensile testing, the elongation of the specimen at failure is of interest. Insufficient elongation will prevent the fabric from conforming to natural conditions, while excessive elongation will allow for plastic yield and inadequate stress mobilization. The required value is the measured elongation divided by the original length between measuring points. ASTM method D 1682 refers to this value as "apparent elongation" due to potential slippage of the fabric in the loading jaws. Some fabric specifications prescribe a minimum fabric toughness, which is the tensile strength times the elongation or the area under the stress-strain curve.

CREEP BEHAVIOR. ASTM method D 2990 gives a test procedure for fabric creep behavior. However, the situation is very much in the research stage. Finnigan[9] has tested polyesters and polyamides, finding that both deform linearly with respect to log time and can be used in the prediction of creep levels. These tests varied from 1 minute to 1,000 hours. Other data, such as, from Monsanto,[10] shows that polypropylene fiber has a markedly different behavior than nylon or polyester fibers. This suggests that the fundamental behavior of synthetic fibers cannot be easily generalized. The interaction of fibers in the form of fabrics made by different materials and processes might also behave quite differently.

TRANSVERSE ELASTICITY TEST. This test, developed by duPont,[11] stretches the fabric over the end of a 6-inch-diameter pipe and pushes a 2-inch square rod into it in a transverse direction. The stress developed in the sample after 1 inch of a penetration is recorded for a modulus value. The operation is repeated fifty times to evaluate the fatigue characteristics of the fabric.

INDIRECT TENSILE TEST. This similar and newly devised test to indirectly measure the tensile strength of a fully constrained specimen has been proposed by Alfheim and Sorlie.[12] The specimen is horizontally clamped in a 6-inch-diameter California Bearing Ratio (CBR) mold,[13] and a 2-inch-diameter piston is pressed into it at a constant speed of 1.3 millimeters per minute until failure occurs. The tensile strength and elongation are straightforward to compute. It is a form of the transverse rupture test.

TRAPEZOIDAL TEAR TEST. This test, ASTM D 2263, now discontinued, was originally devised for testing of automotive fabrics. The trapezoidal tearing load is the force required to successfully break individual yarns (fibers) in a fabric. In this test, the fabric is inserted in a tensile testing machine on the bias so that the fibers are caused to tear progressively. An initial $\frac{5}{8}$-inch cut is made to start the process. The load actually stresses the individual fibers gripped in the clamps rather than stressing the fabric structure. The test was discontinued by ASTM in 1976 without replacement.

TONGUE TEAR. This test, ASTM D 751, uses a 3 by 8-inch fabric specimen with a precut 3-in-long initiation cut. The fabric is placed in a testing machine with the cut ends in the jaws of the machine. An increasing force is applied to make the fabric tear along the initiation cut. The test configuration permits the yarns to "rope up" and work together to resist tear propogation. Thus tongue tear test results are usually much higher than results from trapezoidal tear tests.

IMPACT STRENGTH (ELMENDORF TEAR). A critical mechanical property in some construction applications is impact strength. The ASTM D 1424 test covers a procedure for the determination of the average force required to propagate a single-rip tongue-type tear starting from a premade cut in a woven fabric. The cut is then continued by means of a falling pendulum apparatus. The tearing force is the force required to continue the tear previously started in the specimen. The strength is calculated as the work done in tearing the specimen divided by twice the length of the tear. The test is often used exclusively in Europe to measure tear strength but has questionable validity for nonwoven fabrics.

BURST TEARING (MULLEN BURST). Out-of-plane loading is a common situation in construction use of fabrics, and the test sometimes used to simulate this mechanical property is ASTM method D 774. In this test an inflated rubber membrane is used to distort the fabric into the shape of a hemisphere. Bursting occurs when no further deformation is possible. Thus it indirectly measures both fracture and deformation, although the stress state is very complex.

CONE PENETRATION IMPACT TEST. Alfheim and Sorlie[12] have introduced a new test in which a cone is dropped from a specified height into a horizontally held fabric specimen contained in a 6-inch-diameter California bearing ratio (CBR) mold. The size of the resulting hole is measured and qualitatively rated from one fabric to another. The test attempts to simulate the dropping of stone subbase directly on fabric. Though not an ideal test, it typifies attempts being made to assess impact resistance of construction fabrics.

ABRASION RESISTANCE OF FABRICS. The ASTM test methods for abrasion resistance of textiles fabrics are designated D 1175 and cover six different procedures:

- Inflated diaphragm.
- Flexing and abrasion.
- Oscillatory cylinder.
- Rotary platform, double head.
- Uniform abrasion.
- Impeller tumble.

In all cases, abrasion is defined as "the wearing away of any part of a material by rubbing against another surface." (ASTM Test method D 1175). There are, obviously, a large number of variables to be considered in such a test. Results are reported as the percent weight loss under the specified test and its particular conditions.

FATIGUE. Celanese[14] has modified the ASTM Grab Tensile Test (D 1682) to give a fatigue resistance value under their test designation CFMC-FEET-6. Fatigue resistance is defined as the ability to withstand repetitive loading before undergoing catastrophic failure.

The specimen is stressed longitudinally at a constant rate of extension to a predetermined length, then back to zero load. This cycling is repeated until failure occurs. Eight-inch-long specimens, four inches wide, are tested on an Instron fatigue testing machine. The resulting stress-strain response can be used to calculate a cyclic secant modulus that becomes evident after a number of load cycles are applied.

Although many variables remain to be defined (primarily, the decision as to what deformation to use during testing), the test does simulate *in situ* conditions better than most static tests.

2.3.3 Tests for Hydraulic Properties

POROSITY OR OPENING SIZE OF FABRIC. An important property insofar as drainage through fabrics is concerned is the porosity or size of opening of the fabric. This is most accurately determined by tests common to the area of quantitative metallography. Such devices as "image analyzers," which rapidly scan the specimen and give, by light reflection, the percentage of void to fiber, are ideal for this use.[15] Most systems also have data storage capabilities which allow for scanning of a relatively large specimen and thereby obtain meaningful statistical data on the mean size of openings and the standard deviation or variance of these openings. The disadvantages are the high initial cost of the equipment and the long sample preparation time.

As an alternative, an equivalent opening size (EOS) test has been devised by the Corps of Engineers.[16] The EOS is defined in CW-02215 as the number of the U. S. Standard sieve that has openings closest in size to the filter fabric openings. The test uses known sized glass beads of designated EOS number and determines by sieving (using successively coarser fractions) that size of beads for which 5 percent or less pass through the fabric. The EOS of the fabric sample is the "retained on" U. S. Standard sieve number of this fraction. The test is similar to Celanese' CFMC FEET-1 (Ref. 14), with the exception of the duration of shaking time. Typical values for selected fabrics are given in Table 2.3.

AIR PERMEABILITY. ASTM method D 737 defines air permeability as "the rate of air flow through a material under a differential pressure between the two fabric surfaces." It is expressed as cubic feet per minute of air per square foot of fabric between the stated differential pressures.

TABLE 2.3. Corps of Engineers Data on EOS for Various Fabrics

Fabric	EOS Sieve Number
Filter X	100
Polyfilter GB	40
Polyfilter X	70
Laurel erosion control cloth	100
Mirafi 140	100
Monsanto E2B	80
Nicolon 66487	30
Nicolon 66429	40
Nicolon 66424	50

Source: Ref. 16.

Pressures equivalent to 0.5 inch (12.7 millimeters) of water are often used; however, higher pressures can be selected, depending on the fabric. Obviously, the pressures must be stated along with the test results. The air flow is taken from the test equipment manufacturer's calibration charts for a given air pressure level (as determined from an oil controlled vertical manometer) and air outlet nozzle size.

WATER PERMEABILITY. While not a formal test, it seems logical to model the water permeability of a fabric to water permeability in soil. New York State[17] has proposed such an adaptation of a constant head permeability test.

A specimen of the fabric to be tested is placed on a plastic base plate and fitted with another plastic tube ($4\frac{1}{2}$ inches in diameter) on top of it. Water is allowed to enter the upper tube, from where it then flows down through the fabric and out of the system. A constant head situation is maintained. Three tests are conducted at heads of 3, 12 and 36 inches, respectively. Data are plotted in the conventional manner[13] to obtain the fabric permeability in centimeters per second. Note that this unit requires an estimate, which has not been standardized for construction fabrics, of the fabric thickness. Alternatively, the test can be run with 4 to 5 inches of head and the flow rate measured in gallons/minute per square foot, as suggested by Parks.[18] duPont also uses a constant head test with the permeability measured at 10 inches of head.[11]

The Celanese test[14] utilizes a falling (variable) head permeability procedure. Their recommended procedure follows along standard geotechnical testing practices[13] and uses a 1-inch-diameter cross-sectional area of flow.

PLANAR WATER FLOW (SYPHONAGE). Although not a standard test, the ability of a fabric to carry water horizontally in its manufactured plane is often very important. This type of planar permeability can easily be modeled at different heads and even for negative heads. This latter suggestion would then be measuring the syphonage ability of the fabric. Of importance in such measurements would be the transverse load applied to the fabric as it is under test. Syphonage ability of fabrics has been shown by some manufacturers,[10] but no attempt at test standardization has been made, to our knowledge.

GRADIENT RATIO TEST. This test, designated CW-02215 by the Corps of Engineers,[16] measures gradient ratio and as such is also an indicator of blinding or clogging of the fabric. The gradient ratio is defined as the ratio of the seepage gradient through the fabric and 1 inch of soil to the gradient through an adjacent 2 inches of soil.

The test is performed in a constant head permeameter[13] with the fabric clamped firmly within it and the soil placed above it. Water flows downward, through the soil, then through the filter and finally out of the permeameter. The water is run for 24 hours before data are taken to calculate the gradient ratio. However, by having the test run for considerable lengths of time and observing changes in flow (and/or head variations) an indication of the blinding or clogging of the fabric can be made.

UPWARD GRADIENT TEST. By reversing the usual flow direction of constant head permeability testing one can create an upward hydraulic gradient and thereby assess the fabric's performance in this mode. One can also observe if, and how much of, a "cake" is formed beneath the fabric. The test configuration has an upward flow through the soil and then through the fabric. In order to better simulate natural conditions, a gravel layer can be placed on top of the fabric. Work is currently ongoing at Drexel University[19] and at Celanese[20] in this regard. Since the containment vessel is of plastic, one can visually observe if a stratified condition is forming in the soil beneath the fabric.

The testing proceeds in the usual manner and the fabric/soil system permeability is compared to the permeability of the soil without fabric to note the influence of the fabric. As with the gradient ratio test, the variation of permeability with time of the fabric/soil system will be indicative of the amount of blinding or clogging that the fabric may be undergoing.

FILTER FABRIC SOIL RETENTION TEST. The New York State DOT[21] has devised a laboratory test to evaluate the ability of filter fabrics to act as silt curtains or fences. The test apparatus consists of two plastic tanks, each measuring 18 inches long by 12 inches high by 5 inches wide, with one end of each tank being open and capable of holding the fabric to be tested in a fixed and vertical position. One tank is filled with water and given a dry weight of soil from the site of the proposed construction. After a prescribed mixing procedure the soil-water suspension is allowed to pass through the fabric separating the two tanks. Hydrometer analyses[13] are used to evaluate the weight of soil in each tank and thereby the percent of soil passing the fabric. Head differentials allow for calculation of the velocities involved for repeated trials.

The results of the tests produce data of the percentage of soil passing through a given fabric at various flow velocities. This can be plotted and different fabrics compared.

FABRIC PIPING PERFORMANCE. If flow gradients from the soil become too large for the fabric to retain them, failure of the fabric will result. Using a laboratory test setup similar to the previously described upward gradient test, Celanese[14] localizes flow so as to generate these high gradients. Piezometric levels are measured throughout the soil column beneath the fabric and on both sides of the fabric. Failure of the fabric occurs in the form of a "blow out," with underlying soil flowing upward through the failed fabric.

2.3.4 Tests for Endurance and Miscellaneous Properties

RESISTANCE TO CHEMICAL REAGENTS. ASTM method D 543 covers this area under the title "Resistance of Plastics to Chemical Reagents." The test method includes provisions for reporting changes in weight, dimensions, appearance, and strength properties. Provisions are also made for various exposure times and exposure to reagents at elevated temperatures. A list of fifty standard reagents is supplied in order to attempt some sort of standardization.

For example, duPont[22] has evaluated most of its fibers (acetate, dacron, nylon, orlon, rayon, cotton, wool, silk, etc.) under a wide range of chemicals (sulfuric acid, hydrochloric acid, nitric acid, hydrofluoric acid, phosphoric acid, organic acids, sodium hydroxide, bleaching agents, scouring and laundering agents, salt solutions, organic and miscellaneous chemicals), many of which were at different concentrations and at different temperatures. After the specified exposure, the samples were rinsed, air dried, and then conditioned at 70°F and 65 percent relative humidity for 16 hours.

Data on breaking strength, breaking elongation, and toughness of the scoured fibers were compared to control specimens of the fibers that were not exposed to the chemical.

RESISTANCE TO LIGHT AND WEATHER. This test is also covered by ASTM under the title, "Outdoor Weathering of Plastics," and is designated D 1435. It is intended to define conditions for the exposure of plastic materials to weather. It is a comparative test depending upon climate, time of year, atmospheric conditions, and so on, and, as such, gives only an indication of the long-term *in situ* behavior.

Aluminum racks are constructed with the fabric to be tested fixed to them. Test specimens can be placed at 0°, 45°, or 90° to the horizontal and in different solar orientations. Exposure test samples should simulate service conditions of the end use condition as far as practical.

Most fabric manufacturers will provide information on their products regarding light and weather resistance. For example, DuPont[23] has evaluated acetate, polyester, nylon 6-6, spandex, acrylic, rayon, glass, nylon 6, olefin (polypropylene), rubber, saran, cotton, flax, silt, and wool in Florida for periods ranging from 1 to 36 months. Both direct and "under glass" tests were conducted. In all, fifty different fibers representing fourteen different generic classes were investigated. Breaking strength was the basic property measured and comparisons were made before and after exposure.

It is important in the use of construction fabrics to avoid and/or minimize exposure to ultraviolet light, since such exposure can cause rapid degradation of strength. Polypropylene, if untreated, can be particularly sensitive in this regard.

RESISTANCE TO TEMPERATURE. ASTM Recommended Practice D 794 describes high-temperature testing for plastics. Only the procedure for heat exposure is specified, the test method being governed by the potential end use. Heat is applied using an oven with controlled air flow and with substantial fresh air intake. Two types of tests are described: continuous heat and cyclic heat. In the former, heat is gradually increased until failure occurs. Failure is defined as a change in appearance, weight, dimension, or other property that alters the material to a degree that it is no longer serviceable for the purpose in question. The test may take minutes to weeks, depending on the rate of temperature increase. The cyclic heat test repeatedly applies heat of a constant value until failure.

Caution in the use of high temperatures is to be exercised when placing hot materials such as asphalt or joint sealer on construction fabrics because of their high-temperature sensitivity.

ASTM method D 746 addresses the effect of cold temperature on plastics and, in particular, the properties of brittleness and impact strength. At various temperatures, specimens are tested by a specified impact device in a cantilever beam test mode. Brittleness is defined as "that temperature estimated statistically, at which 50 percent of the specimens would fail in the specified test." Numerous samples are required for testing, since a statistical value is required.

BURIAL DETERIORATION OF FABRICS. The National Research Council of Canada[24] has tested the effects of burial on fabrics, thereby recognizing that soil is very variable material. It ranges from $\simeq$99 percent organic to 100 percent inorganic, has a wide range of pH values, and varies greatly in elemental compositions and microorganism contents. The test involves 12 by 12 centimeter fabric samples of polyethylene terephthalate, polypropylene, and nylon-polypropylene bicomponent fabrics. This test method is designated CGSB 4-GP-2 method 28.3 and is similar to AATCC test method 30-1974 and Federal Standard No. 191, method 5762.

Samples are removed at three-month intervals and are tested according to the diaphragm pressure (Mullen burst) test found in ASTM method D 774. Future testing will involve other fabrics and a wide range of soil conditions.

2.4 Overview of Construction Fabrics

Of paramount interest for the purposes of this book are the trademarks of various formed fibers. See Appendix 2 for other trademarks and note that this is by no means a complete listing. Table 2.4 lists the materials that comprise the majority of fabrics used in geotechnical construction. They are individually described in detail in Section 2.5.

2.5 Details of Construction Fabrics

In this section details of commonly used fabrics in geotechnical construction are presented. All properties presented (physical, mechanical, hydraulic, and environmental) are taken directly from manufacturers' literature. As such, great variations in tests, test procedures, and methods of reporting data will be noted. Most model numbers, styles, data, and so on reflect currently (1979) available information.

The products are not limited to only American or North American manufacturers; many overseas products are also listed. By no means is the listing complete since the field is in a rapid state of change with new firms entering (and some leaving) on a continuing basis. The section numbers are keyed to the numbered fabric trademarks presented in Table 2.4.

Note should also be made that manufacturers often change specifications of their fabrics while retaining the same style classifications. The

TABLE 2.4. Trademarks of Various Formed Fibers (Fabrics)

Section Number	Trademark	Fiber & Process	Manufacturer or Sales Agent
2.5.1	Adva-Felt	Polypropylene, nonwoven	Advance
2.5.2	Bay Mills	Glass, woven	Bay Mills Midland
2.5.3	Bidim	Polyester, spun bonded and needled	Monsanto
2.5.4	Cerex	Nylon, spun bonded	Monsanto
2.5.5	Cordura	Nylon, woven	duPont
2.5.6	Enkamat	Nylon, melt bonded	American Enka
2.5.7	Fibertex	Polypropylene, spun bonded and needled	Crown Zellerback
2.5.8	Filter-X	Polyvinylidene chloride, woven	Carthage Mills
2.5.9	Laurel Cloth	Polypropylene, woven	Laurel Plastics/Advance
2.5.10	Mirafi	Polypropylene and others, nonwoven and woven	Celanese
2.5.11	Monofelt	Polypropylene, nonwoven	Menardi Southern/J. P. Stevens Co.
2.5.12	Monofilter	Polypropylene, woven	Menardi Southern/J. P. Stevens Co.
2.5.13	Nicolon	Polyamide and others, woven and nonwoven	U. S. Textures Sales
2.5.14	Permealiner	Polypropylene, woven	Staff Industries
2.5.15	Petromat	Polypropylene, needle bonded	Phillips Fibers
2.5.16	Polyfelt	Polyester, needled	Advance
2.5.17	Poly-Filter	Polypropylene, woven	Carthage Mills
2.5.18	ProPex	Polypropylene, woven	Amoco Fabrics
2.5.19	Reemay	Polyester, spun bonded	duPont
2.5.20	Sontara	Polyester, spun laced	duPont
2.5.21	Stabilenka	Polyester	American Enka
2.5.22	Supac	Polypropylene, needlebonded	Phillips Fibers
2.5.23	Terrafix	Polyesters and others, nonwoven	Erosion Control
2.5.24	Terram	Polypropylene and polyethylene, thermally bonded	ICI Fibers
2.5.25	Typar	Polypropylene, spun bonded	duPont
2.5.26	Tyvek	Polyethylene, spun bonded	duPont

information can be verified by contacting the manufacturer directly. The addresses are provided in Appendix 4.2.

2.5.1 Adva-Felt

Adva-Felt is a nonwoven polypropylene fabric manufactured by the Docan process available through Advance Construction Specialties Co., Inc.[25] Its intended fields of application are as separation, reinforcement, drainage, and erosion control. As Table 2.5 indicates, the fabric is available in three styles.

TABLE 2.5. Properties of Adva-Felt

	Style Number			
Property	TS 200	TS 300	TS 400	Test Method
Weight (oz/yd^2)	6	8	10.5	ASTM-D 461
Thickness (in.)	0.08	0.13	0.17	"
Tensile strength (lb)				
Dry	115	225	300	ASTM-D 1682-59T
Wet	160	270	420	"
-50°C wet	116	228	308	"
Elongations (%)				
Dry	89	101	110	"
Wet	Same	Same	Same	"
Frozen -50°C	48	51	59	"
Tear resistance, strip (lb/cm)	34	47	64	ASTM-D 2261-62
Biological or marine growth	None	None	None	C. of E.[a]
Permeability (cm/sec)	4×10^{-2}	5.7×10^{-2}	7.9×10^{-2}	C. of E.
Specific gravity	0.91	0.95	0.95	
Equivalent opening size	90	70	60	C. of E.

Source: Ref. 25.
[a] Corps of Engineers.

2.5.2 *Bay Mills*

Bay Mills[26] produces a line of fabrics woven of continuous filament glass yarns for the following general purposes:

- To reinforce gel coats.
- To give a smooth tailored appearance to the inside of parts such as boats.
- To cover wood in boats, furniture, and so on.
- As the main reinforcement in high strength laminates.
- To make molds and tooling in applications such as dies, gages, templates, workholders, drill shell, tooling masters, mockups, patterns, blue blocks, and drop hammer dies.

These fabrics have not seen wide use in civil engineering construction, although they have been considered, that is, laboratory tested, for a number of applications. The fabrics are available in four different styles, as Table 2.6 indicates.

TABLE 2.6. Properties of Bay Mills Fabrics[a]

Style Number	Thread Count (per inch)		Average Tensile (lb/inch)		Weight (oz/yd^2)	Thickness (mils)	Weave	Approximate Yards per Roll
	Warp	Fill	Warp	Fill				
154	16	14	450	410	9.7	14	Plain	130
144	18	18	250	220	6.0	9	Plain	200
610	32	28	115	100	2.48	4	Plain	500
196	28	42	1200	700	32	45	Special	40

Source: Ref. 26.

[a]Note that woven fabrics consist of two perpendicular sets of interlacing fibers. The warp runs lengthwise and can be hundreds or thousands of yards long. The fill, or interlacing yarn, is made to the width of the fabric.

2.5.3 *Bidim*

Bidim® engineering fabrics are made by the Monsanto Company[10] from continuous filament polyester fibers that are spun and then mechanically entangled by needle punching. The fabrics are supplied in a choice of five different styles, depending on the type of end use, the soil condition, and the service demand on the fabric. These styles are designated C22, C28, C34, C38, and C42 in order of increasing thickness. Current uses are in separation, reinforcement, drainage, erosion control, and other construction applications. Table 2.7 shows typical values of the physical properties of the various styles.

TABLE 2.7. Bidim® Engineering Fabrics[a]

	Style Number				
Property[b]	C22	C28	C34	C38	C42
Weight/roll (lb)					
13 ft 10 in. wide	425	565	760	950	760
17 ft 5 in. wide	535	715	950	1200	950
Roll length, (ft)	984	984	984	984	492
Roll diameter (in.)	40	40	40	40	40
Thickness, ASTM D-1777 (mils)	60	75	90	110	190
Grab tensile strength (lb force)	115	160	255	300	610
Grab elongation (%)	85	80	75	65	60
Trapezoid tear strength (lb force)	62	93	125	170	250
Mullen burst strength (psi)	225	360	400	500	850
Restrained tensile test (lb force/in.)[c]	65	95	120	150	270
Elongation (%)	35	35	35	35	35
Normal permeability (10^{-3} m/sec)[c]	3	3	3	3	3
Planar permeability (10^{-3} m/sec)[c]	0.6	0.6	0.6	0.6	0.6
Equivalent opening size					
D_5	50	50	70	100	100
D_{50}	70	100	100	140	140
Abrasion resistance (grab strength) (lb)[d]	40	120	135	165	295
Heat resistance @ 50 psi loading (°F)[c]	480	480	480	480	480
pH resistance range (pH)			3 to 11		
Puncture strength ASTM D-751—modified (lb force)	55	95	125	145	255
Porosity (%)	93	92	91	91	91

Source: Ref. 10.

[a]Physical properties data represent typical values and should not be construed as absolute values or specifications.

[b]Test methods ASTM D-1117-69, D-1682, and D-2263.

[c]Monsanto test.

[d]Modified Corps of Engineers tests.

2.5.4 Cerex

Cerex® fabrics[10] are made from nylon 6.6 and are spun bonded and manufactured by the Monsanto Company. They are nonwoven fabrics made directly from the molten polymer as continuous filaments that are self-bonded at each crossover point by a proprietary process. The fabrics contain no adhesive or other additives. The main thrust of the fabric has not been the construction market, but it has been used in the drainage area in protection of porous pipe underdrains. The fabric is produced in 9 different weights from 0.3 to 2.0 ounces per square yard, as Table 2.8 indicates.

2.5.5 Cordura

Cordura® is duPont's registered trademark for its high-tenacity nylon fiber.[11] It is a producer-bulked industrial filament yarn in which the filaments have been disarranged, looped, and tangled within the yarn bundle. Being a filament nylon fiber with high bulk characteristics, it is used to make industrial broadwoven fabrics and webbings. It is used in the context of construction fabrics to make flexible fabric forms, as described in Chapter 7. Other construction uses described in this book could also make use of this material. Table 2.9 presents the properties of two types of woven fabrics made from Cordura fiber.

2.5.6 Enkamat

Enkamat® is a three-dimensional soil reinforcement matting of very open construction made from heavy nylon monofilaments bonded at their intersections. Patents are pending on this product of the American Enka Company, a part of Akzona Incorporated, Enka, North Carolina.[27] Applications include ditch lining, slope stabilization, and lakeshore and riverbank erosion situations. Table 2.10 gives the specifications of the two types of Enkamat Matting currently available.

An interesting combination of Enkamat and Stabilenka, both products of the American Enka Company, is Enkadrain. Enkadrain is a two-layer composite of the two different fabrics, where the Stabilenka is bonded to the thicker and more porous Enkamat. The combination prevents soil from entering the system and allows for extremely high in-plane flow. It is designed for relieving hydrostatic pressure adjacent to underground basement walls and retaining walls. Enkadrain is available in a 38.2-inch width plus a 3-inch filter fabric overlap, in a length of 30 meters per roll,

TABLE 2.8. **Physical Properties of Cerex® Fabrics**[a]

Fabric Weight[b] (oz/yd^2)	Average Thickness[c] (mils)	Grab Strength[d] (lb)		Tear Strength[e] (lb)		Mullen Burst[f] (psi)	Air Permeability[g] [$(cm/ft)/ft^2$]
		Machine Direction	Transverse Direction	Machine Direction	Transverse Direction		
0.3	2.3	8	5	3.4	2.4	17	1300
0.4	2.5	12	7	4.3	3.0	20	1050
0.5	3.2	16	11	4.5	3.5	24	850
0.6	3.4	21	15	5.5	4.5	29	700
0.7	3.7	27	18	6.3	5.3	33	600
0.85	4.2	32	21	6.5	5.4	36	470
1.0	4.8	41	26	8.0	6.7	40	360
1.5	7.1	53	40	11.0	10	52	220
2.0	8.7	70	54	14.0	13	65	160

Source: Ref. 10.

[a]These data are representative of Cerex fabrics and are not intended to serve as specification. Detailed testing procedures are available from Monsanto on request.

[b]ASTM D-1910-64.

[c]ASTM D-1777

[d]ASTM D-1682

[e]ASTM D-2263-68

[f]ASTM D-231

[g]ASTM D-737-69

TABLE 2.9. Properties of Fabrics Made from Cordua® Fibers

Properties	500-Denier Cordura	1,000-Denier Cordura
Weight (oz/yd^2)	6.3	9.9
Thickness (in.)	0.017	0.027
Bulk (cm/g)	2.0	2.1
Grab strength–warp (lb)	413	567
Tounge tear strength–warp (lb)	34	47
Mullen burst strength (psi)	495	680
Modified Wyzenbeek abrasion–warp (cycles to failure)	1273	1463

Source: Ref. 11.

TABLE 2.10. Specifications of Enkamat® Matting

Property	Style Number	
	7010	7020
Material		
Nylon 6 plus a minimum content of 0.5% by weight of carbon black		
Dimensional		
Weight (g/m^2)	265 ± 7%	405 ± 7%
Thickness, minimum (mm)	9	18
Width (cm)	97 ± 3%	97 ± 3%
Roll length (m)	150 ± 3	100 ± 3
Filament diameter, minimum (mm)	0.35	0.40
Tensile strength, minimum (kg/m)[a]		
Length direction	80	140
Width direction	40	80
Tensile Elongation, minimum (%)[a]		
Length direction	50	50
Width direction	50	50
Tensile Resiliency		
30 minute recovery (%) (3 cycles at 100 psi)	80	80
Exposure properties		
Temperature range for 80% strength retention (°F)	-100-250	
pH range for 80% strength retention	3-12	

Source: Ref. 27.

[a]ASTM 1682 strip test procedure modified to obtain filament bond strength is used to indicate tensile properties of Enkamat matting.

with a roll diameter of 38 inches, thickness of 0.8 inch, and a total weight of 715 grams per square meter.

2.5.7 *Fibretex*

Fibretex® is a nonwoven needle punched spun-bonded polypropylene fabric produced by Crown Zellerback[28] and called Construction Grade Fibretex. It is used in separation, reinforcement, drainage, erosion control, forms, and has other construction applications. There are currently six grades available, as Table 2.11 indicates.

TABLE 2.11. Properties of Fibretex® Fabric

	Grade		
Property	320	420	600
Weight (g/m^2)	320	420	600
Strip tensile (lb/in.)	75	100	150
Elongation (%)	130	150	160
Rupture energy (ft-lb/ft^2)	800	1,000	2,000
Grab tensile (lb/in.)	125	150	250
	Grade		
Property	200	300	400
Weight (g/m^2)	200	300	400
Tensile strength (lb/in.)	60	85	100
Elongation (%)	85	130	150
Permeability (cm/sec)	0.19	0.18	0.17

Source: Ref. 28.

2.5.8 *Filter-X*

Filter-X® is a registered trademark of a fabric produced by Carthage Mills[29] and consists of a pervious sheet of polyvinylidene chloride monofilament yarns. The yarn consists of at least 85 percent vinylidene chloride and contains stabilizers to make the filament resistant to ultraviolet and heat deterioration. After weaving, the fabric is calendered so that the filaments retain their relative positions with respect to one another. All edges are selvaged or serged. The fabric has been used primarily in silt curtain drainage and erosion applications and has also been used in separation, reinforcement, and other construction problem areas. Table 2.12 gives the pertinent product data.

TABLE 2.12. Filter-X® Product Data

Test	Method	Result
Breaking load and elongation	ASTM D 1682, grab test method, 1 in.-square jaws, constant rate of travel 12 in./min.	Tensile strength: stronger principal direction 200 lb; weaker principal direction 110 lb; elongation at failure between 10 and 35%
Oxygen pressure	CRD-C 577 or 7111 in Fed. Std. 601	Same as breaking load and elongation result
Effect of alkalies	Special	Tensile strength: stronger principal direction 190 lb; weaker principal direction 105 lb; elongation at failure between 10 and 35%
Effects of acids	Special	Same as breaking load and elongation result
Weight change in water	CRD-C 575 or 6631 in Fed. Std. 601	Less than 1%
Brittleness	CRD-C 575 or 6631 in Fed. Std. 601	No failure at -60°F
Freeze-thaw	CRD-C, modified	Tensile strength: stronger principal direction 195 lb; weaker principal direction 105 lb; elongation at failure between 10 and 35%
Bursting strength	ASTM D 751	260 lb/in.^{2}
Puncture strength	ASTM D 751, modified	70 lb
Seam breaking strength	ASTM D 1683	80 lb
Percent of open area	Special	4–5%
Equivalent opening size	Special	U.S. Std. Sieve #100 (0.149 mm)
Permeability	—	4.8×10^{-2} cm/sec
Specific gravity	—	1.70
Weight	—	$\simeq$0.833 lb/ft^2 $\simeq$1.29 oz/yd^2

Source: Ref. 29.

2.5.9 *Laurel Cloth*

Laurel Erosion Control Cloth (LECC) is a woven polypropylene monofilament yarn made by Laurel Plastics Co., for Advance Construction Specialties Co.[25] It has been used primarily to protect against erosion of subgrade materials in water control structures and in highway and retaining system drainage cases.

The fabric is available in two different styles. Table 2.13 gives test results for the two types.

TABLE 2.13. Laurel Erosion Control Cloth Test Results

Property	Type A (I)	Type B (II)	Test Methods
Thickness (mils)	17	22	—
Weight (oz/ft^2)	0.80	0.70	ASTM D 1910
Equivalent opening size (EOS)	100 U.S. Std. sieve	40 U.S. Std. sieve	C. of E.
Open area (%)	4.3	26	C. of E.
Break strength, grab (lb)			ASTM D 1682-64
Warp	399	280.2	
Fill	244	232.2	
Elongation (%)			ASTM D 1682-64
Warp	33	40.2	
Fill	33	42.4	
Trapezoidal tear strength (lb)			ASTM D 1682-64
Warp	90	—	
Fill	35	—	
Burst strength, Mullen (psi)	528	520	ASTM D 751-68
Puncture (lb)	138	133	ASTM D 751-68
Loss of strength when wet	Nil	Nil	C. of E.
Abrasion, Taber cycles	No damage at 3700 cycles	No damage at 3700 cycles	ASTM D 1175-64T D 1682-64
Seam strength (lb)	198	198	ASTM D 1683-68
Effect of salt water	Nil	Nil	
Weatherometer test, tensile strength (%)			
Warp	87	88	ASTM G 23 &
Fill	74	73	D 1682-64
Moisture absorbancy	Nil	Nil	CRD-C 575
Biological or marine growth	None	None	C. of E.
Specific gravity	0.95	0.95	

Source: Ref. 25.

2.5.10 Mirafi

Mirafi is a licensed trademark of the Celanese Corporation.[14] Mirafi 140® fabric is constructed from two types of continuous filament fibers. One is polypropylene, and the other is a heterofilament comprised of a polypropylene core covered with a nylon sheath. A random mixture of these filaments is formed into a sheet that is heat bonded, with the result that the heterofilaments are directly fused at their points of contact. The polypropylene filaments remain unaffected during the heat-bonding process and are held within the matrix by purely mechanical links. Mirafi 140 is marketed by the Celanese Corporation. Past uses are in separation, reinforcement, drainage, erosion, and other construction areas.

A recently introduced fabric of Celanese is Mirafi 500X®. It is a fabric woven from monofilaments of isotatic polypropylene. The fabric is then heat treated and the edges are mechanically sealed to prevent unraveling. Also recently introduced by Celanese is Mirafi 100X®, a woven fabric constructed from polypropylene fibers for applications as silt fences and brush barriers. The fabric is available in widths of 36 and 48 inches.

Table 2.14 gives typical properties of these three Mirafi fabrics. These are average values and should not be construed as minimum specification values.

2.5.11 Monofelt

Monofelt™ is a nonwoven polypropylene fabric marketed by Menardi Southern and produced by J. P. Stevens Co., Inc.[30] by a special process that entangles and fuses the individual fibers. Although it is specifically engineered for the erosion control market, it can be used for separation, reinforcement, and drainage applications in construction problems. Table 2.15 gives some of its important properties.

2.5.12 Monofilter

Monofilter® fabric is woven entirely from monofilament polypropylene yarns and then put through a specific finishing process by Menardi-Southern under the trademark of J. P. Stevens Co., Inc.[30] The yarns used in production of the fabric have been treated with ultraviolet light inhibitors for resistance to this type of degradation. Its applications have been as follows:

- Filters for shore protection and coastal structures.
- Road and area stabilization.

TABLE 2.14. Mirafi® Product Information Summary

Property	Fabric		
	Mirafi 140	Mirafi 500X	Mirafi 100X
Minimum weight (g/m^2; oz/yd^2)	140; 4.1	136; 4.0	–
Average thickness (mils)	30	25	–
Grab strength, wet (lb)	120	200	120
Retention at -70°F (%)	100	–	>90
Grab elongation, wet (%)	130	–	–
Retention at -70°F (%)	40	–	–
Burst Strength (Mullen) (psi)	–	325	200
Tear strength (lb)	65[a]	25[b]	65[a]
Air permeability	250	–	–
Fabric width	14 ft 9 in. (4.5 m)	12 ft 6 in. (3.8 m)	36 or 48 in.
Length per roll	328 ft (100 m)	430 ft (131 m)	300 or 225 yd
Average weight per roll (lb)	170	180	55 or 60
Equivalent opening size			40

Source: Ref. 14.
[a]Trapezoidal
[b]Elmendorf.

TABLE 2.15. Typical Properties of Monofelt™ Fabric

Item	Properties	Test Method
Fiber	100% Polypropylene	
Weight/yd^2 (oz)	5	ASTM: D1910-64, 32
Grab tensile, MD X CMD (lb)	Original: 130 X 165	ASTM: D1682
	After abrading: 100 X 112[a]	ASTM: D1682[a]
Elongation, MD X CMD (%)	72 X 64	ASTM: D1682
Mullen burst (psi)	230	ASTM: D1117-62
Equivalent opening size	80–100	U.S. Army Corps of Engineers CW-02215–Oct. 1976

Source: Ref. 30.
[a]After abrading, as in D1175, with CS17 Celibrase wheels under 1 kg load; 1,000 revolutions.

TABLE 2.16. Fabric Specifications for Monofilter® Fabric: yarn—100% polypropylene monofilament black; weight—7 oz/yd^2 +; thickness—0.020 in.

Property	Style Number 17-980-040-72	CW-02215
Tensile strength (ASTM D-1682—grab method)		
Warp	347	200 lb—any direction
Fill	332	
Bursting strength (ASTM D-751—Mullen method), gross	532	Not specified
Puncture strength (ASTM D-751—ball burst method)	250+	120 lb (minimum)
Abrasion Resistance (ASTM D-1682 after D-1175—Taber abrasion, 1,000 cycles, CS-17 wheel, 1000 g)		
Warp	120	55 lb—any direction
Fill	304	
Open area (%)	26	
Designated EOS	35–40	

Source: Ref. 30.

- River and stream erosion control.
- Drainage systems for highways and slopes.

Table 2.16 presents the fabric specifications.

2.5.13 Nicolon

United States Textures Sales Corporation[31] is the sole supplier of Nicolon filter cloths and patented systems in America. Their fabrics consist of polyamide, polypropylene, and polyester fibers and are in both woven and nonwoven forms. Nicolon fabrics are also manufactured with loops, pockets, and in an impervious form. Application areas over the past 20 years have been mainly concerned with drainage and erosion control problems, but separation and reinforcement problems have also been addressed. In Table 2.17 physical and mechanical properties are presented for ten different Nicolon fabrics.

TABLE 2.17. Properties of Nicolon Fabrics

Property	Fabrics[a]			
North American designation	Nicolon 40	Nicolon 40L[a]	Nicolon 70	Nicolon 70L[a]
European designation	66339 (301)	66373	66424	66392
New fabrics, 1979				
Composition	Polyethylene or polypropylene monofilament	Polyethylene or polypropylene monofilament	Polyethylene or polypropylene monofilament Warp polypropylene multifilament fill	Same as Nicolon 70
Tensile strength				
ASTM D 1682				
grab test (lb/in.)				
Warp	260	240	240	240
Fill	225	210	230	230
ASTM D 1682				
strip test (N5cm)				
Warp	2,340	2,095	2,100	2,125
Fill	2,070	1,860	2,045	2,040
Bursting strength ASTM D 751 (psi)	500	500	>600	>600
Puncture strength, ASTM D 751 modified (lb)	150	120	120	120
Abrasion resistance, ASTM D 1175				
Abraded strength, D 1682 (lb)	115	100	>60	>60
Equivalent opening size				
U.S. standard sieve	40	35	70	70
Sieve size (microns)	420	500	210	210
Elongation percent				
Warp	30	32	30	30
Fill	25	20	40	40
Water permeability as determined by Delft hydraulic laboratory				
K value (cm/sec)	3	3.5	0.6	0.6
Hydraulic gradient	0.49	0.45	1.05	1.05
Weight				
(g/m^2)	220	195	250	240
(oz/m^2)	6.4	5.67	7.25	7
Thickness				
(mils)	30	30	24	24
(mm)	0.79	0.75	0.61	0.61
Standard widths (ft)	12–16.5	16.5	12–16.5	16.5
Principal applications	Hydraulic filter	Sinking mattresses Hydraulic filter	Hydraulic filter	Sinking mattresses Hydraulic filter

TABLE 2.17. (Continued)

Property	Fabrics[a]				
	Nicolon MD7500 66186	Nicolon HD10,000 66475	Nicolon HD20,000	Nicolon LD1,000	Nicolon X	Nicolon HD40,000
	Polyamide (nylon) multifilament	Polypropylene	Polyester multifilament Warp-polyamide multifilament fill	Polypropylene	Polypropylene monofilament	Polyester multifilament Warp-polyamide multifilament fill
	400	1,250		200	>350	
	400	225		200	>200	
	3,735	10,800	200 kN/m			
	3,490	2,030	60 kN/m			
	>600	>1,500		>325	>500	
	120	>200			120	
	60	>200			>80	
	100	35			70–100	
	150	500			210–150	
	30	18	9		30	
	32	18	24		30	
	6×10^{-2}	10×10^{-2}			$>3 \times 3^{-2}$	
	90.5	18.7				
	230	730	450	140	240	
	6.75	21.5	13.3	4	7	
	19.7	89	30		18	
	0.5	2.26	0.8		0.48	
	16.5	16.5	16.5	12	6–12	
	Reinforcement Hydraulic filter	Reinforcement Hydraulic filter	Reinforcement	Separation	Hydraulic filter	

Source: Ref. 31.

[a]Nicolon 40L and 70L are patented filters with loops woven in the fabric at 30-in. intervals to facilitate attachment of fascines.

2.5.14 Permealiner

Permealiner® is a woven, polypropylene fabric marketed by Staff Industries, Inc.[32] It has generally been used in drainage and erosion areas but could be used for separation and reinforcement as well. Table 2.18 gives the manufacturer's data on the two styles of woven polypropylene,

TABLE 2.18. Manufacturer's Data on Permealiner® Fabrics

Property	M-1195	M-1105	Test Method
Color	Black	Black	—
Weight (oz/ft^2)	0.8	0.72	ASTM D-1910
Equivalent opening size	70–100	30	U.S. standard sieve size CW-02215
Open area (%)	4–10	22	CW-02215
Tensile strength (lb)	400 × 280	275 × 300	ASTM D-1682
Elongation (%)	34 × 32	28 × 32	ASTM D-1682
Trapezoidal tear strength (lb)	92 × 40	110 × 80	ASTM D-2263
Mullen burst (psi)	510	520	ASTM D-751
Puncture strength (lb)	150	130	ASTM D-751-M
Abrasion resistance			ASTM D-01175-71
Abraded strength (lb)	80	70 × 120	ASTM D-1682
Weather-Ometer strength retention (%)	90	90	Federal test method CCC-T-191B, Method 5804
Water permeability, water flow rates[a] (ml/min)			
6-in. head	460–520		Canvas Products
8-in. head	620–760		Assoc. International
36-in. head	2610–2790		test method
	ISS-1	**ISS-2**	**Test Method**
Color	Black	Tan	
Count	24 × 12	16 × 15	ASTM D-3348
Thickness (mils)	14	18	
Width	15 ft	57 in.	
Tensile strength (lb)			
Warp	110	160	ASTM D-1682
Fill	70	110	ASTM D-1682
Elongation (%)			
Warp	22	12	ASTM D-1682
Fill	18	11	ASTM D-1682
Tear strength (lb)			
Warp	30	30	ASTM D-2263
Fill	40	35	ASTM D-2263

TABLE 2.18. (Continued)

Property	ISS-1	ISS-2	Test Method
Burst (psi)	300	400	ASTM D-751
Puncture strength (lb)	35	100	ASTM D-751 (modified)
Weight (oz/yd^2)	5.0	7.5	ASTM D-1910
Air permeability [(ft^3/min)/ft^2]	24	25	ASTM D-737-46
Sieve number, finer than	120	100	CE 1310
Water permeability [(ml/sec)/cm^2]			
130–30 cm	15.2	8.5	Falling head
50–30 cm	8.7	4.9	permeability
30–10 cm	4.1		
Equivalent opening size	60	120	
Specific gravity	1.03	0.98	

Source: Ref. 32.
[a]Water flow perpendicular to fabric.

M-1195 and M-1105, and on two other styles of nonwoven needle punched fabric, ISS-1 and ISS-2.

2.5.15 Petromat

Petromat® is a registered trademark of the Philips Petroleum Company.[33] The fabric is a needle punched nonwoven polypropylene fabric made by Philips Fibers Corporation. It has been used primarily as a waterproofing and reinforcement fabric in highway and bridge deck construction and remedial work. Table 2.19 gives some of its engineering properties.

TABLE 2.19. **Properties of Petromat® Fabric**

Property	Typical	Minimum
Weight (oz/yd^2)	4.1	3.6
Tensile Strength, ASTM method D-1682 (lb)	115	90
Elongation at Break, ASTM method D-1682 (%)	65	55
Asphalt retention, Philips procedure (gal/yd^2)	–	0.20
Color	Black	
Width (in.) (other widths available)	75 and 150	
Length per roll (yd)	100	

Source: Ref. 33.

2.5.16 Polyfelt

Polyfelt TS 300 fabric is a needled polyester material originally sold in America by Advance Construction Specialties Company.[25] It is currently made in Austria and marketed in Europe. Emphasis has been on its use in road construction (reinforcement) and in hydraulic construction (drainage). Test results of the fabric are given in Table 2.20.

TABLE 2.20. Test Results of Polyfelt TS 300 Fabric

Property	Value	Test Method
Weight (oz/yd^2)	7.8	ASTM D-461
Thickness (in.)	0.127	ASTM D-461
Grab tensile strength (lb)	228	ASTM 1682-71
Elongation (%)	101	ASTM 1682
Tear strength (lb)	47	ASTM 2261/62
Moisture takeup (%)	730	ASTM D 461-61
Permeability (cm/sec)	0.012	—
Form of supply		
Single rolls:	8 ft 2 in. × 787 feet (6430 ft^2)	
Wider rolls:	Increments of 8 ft 2 in. by lengths up to 787 ft. Standard rolls not to exceed 10,000 ft^2. Rolls sewn together with a serging seam.	
Weight:	7.7 oz/yd^2 (345 lb/single roll)	

Source: Ref. 25.

2.5.17 Poly-Filter

Poly-Filter X® and Poly-Filter GB® are registered trademarks of fabrics produced by Carthage Mills[29] and are made from pervious sheets woven of polypropylene monofilament yarns. The yarns consist of at least 85 percent propylene and contain stabilizers and inhibitors to make the filament resistant to ultraviolet and heat deterioration. After weaving, the cloths are calendered and palmered so that the filaments retain their relative positions with respect to one another. All edges are selvaged and/or serged. The fabrics have seen the most use in drainage and erosion applications, and have also been used in separation, reinforcement, and other applications. Table 2.21 gives some product data for the two types currently available.

TABLE 2.21. Polyfilter® Product Data

Test	Poly-Filter X	Poly-Filter GB	Method
Breaking load and elongation	380 × 220 10 and 35%	200 × 200 10 and 35%	ASTM D 1682, grab test method, 1 in. square jaws, constant rate of travel 12 in. per min.
Effect of acids	350 × 220 10 and 35%	Same as above Same as above	Special
Effects of alkalies	Same as above	Same as above	Special
Weight change in water	Less than 1%	Less than 1%	CRD-C 575 or 6631 in Fed. Std. 601
Brittleness	No failure at -60°F	No failure at -60°F	CRD-C 570 or 5311.1 in Fed. Std. 601
Bursting strength (psi)	540	600	ASTM D 751, using diaphragm bursting tester
Puncture strength (lb)	140	120	ASTM D 751, modified
Seam breaking strength (lb)	195	170	ASTM D 1683
Open area (%)	5-6	21-26.5	Special
Equivalent opening size	#70 U.S. Std. sieve (0.21 mm)	#40 U.S. Std. sieve (0.42 mm)	Special
Permeability (cm/sec)	$(3.3\text{-}3.8) \times 10^{-2}$	–	–
Specific gravity	0.95	0.95	–
Weight			
(lb/ft^2)	≃0.05	≃0.046	–
(oz/ft^2)	≃0.8	≃0.736	–

Source: Ref. 29.

TABLE 2.22. Manufacturer's Properties of ProPex II® Fabric

Property	Value	Test Method
Material	Polypropylene	
Color	Black	
Tensile strength (lb)	275 X 300	ASTM D-1682
Burst strength (psi)	520	ASTM D-751
Weight (oz/yd^2)	6.5	ASTM D-1910
Equivalent opening size	30	U.S. Std. Sieve size CW-02215
Open area (%)	22	CW-02215
Elongation (%)	28 X 32	ASTM D-1682
Puncture strength (lb)	130	ASTM D-751 (modified)
Weather-ometer strength retention (%)	90 X 90	Federal test method CCC-T-191B, method 5804
Abrasion resistance		ASTM D-1175-71
Abraded strength (lb)	70 X 120	ASTM D-1682
Trapezoid tear strength (lb)	110 X 80	ASTM D-2263

Source: Ref. 34.

2.5.18 ProPex

ProPex® is a woven, polypropylene fabric by Amoco Fibrics Company[34]. It has been used as a separator and as a filter fabric. Table 2.22 gives the significant properties.

2.5.19 Reemay

Reemay® spun-bonded polyester from duPont[11] is a sheet product of continuous-filament polyester fibers that are randomly arranged, highly dispersed, and bonded at filament junctions.

While this fabric has not been widely used in geotechnical engineering applications, it does appear suitable for many uses and comes in a wide variety of styles. Table 2.23 gives some of its physical properties.

2.5.20 Sontara

Sontara® spun-laced fabric is a duPont material[11] consisting of polyester staple fibers entangled to form a strong unbonded structure. Because this fabric does not have any resin binders or interfiber bonds, the fibers are free to bend and move past one another as the fabric is flexed, thus providing excellent softness and draping characteristics. The fabric comes in two different styles. Its physical properties are given in Table 2.24.

TABLE 2.23. Typical Physical Property Ranges of Reemay® Fabrics

	Style Number										
Property	2006	2011	2014	2016	2024	2033	2408	2416	2431	2441	2470
Nominal basis weight (oz/yd^2)	0.6	0.73	1.0	1.35	2.1	2.9	1.1	1.5	2.4	2.9	5.8
Thickness (mils)	5–7	6–8	8–10	9–11	11–13	15–17	10–14	12–16	16–20	18–22	30–34
Grab tensile (lb)											
MD	9–12	12–15	16–28	25–36	39–64	52–88	12–19	15–29	35–54	46–66	106–127
XD	7–9	8–13	12–19	19–29	23–50	42–70	7–13	14–21	29–40	39–53	83–112
Tongue tear (lb)											
MD	1.5–1.7	1.2–2.3	1.2–2.1	1.9–4.2	1.1–3.4	2.3–4.7	2.1–3.3	3.3–4.5	4.0–6.0	5–6	8–12
XD	1.5–1.7	1.1–2.3	1.2–2.7	1.7–4.4	1.4–4.0	1.4–6.2	2.0–2.8	3.2–4.4	4–6	5–7	11–14
Mullen burst (lb)	8–15	22–40	25–36	31–61	52–78	78–114	15–26	22–33	38–54	48–68	76–100

Source: Ref. 11.

TABLE 2.24. Typical Physical Properties of Sontara® Spun-Laced Fabrics

Property	Style Number	
	8000	8003
Basic weight, average oz/yd^2	1.2	1.9
(g/m^2)	40.7	64.6
Grab break strength, MD/XD		
(lb)	25/13	40/25
(N)	111/58	178/111
Elongation to break, MD/XD (%)	40/110	40/100
Tongue tear, MD/XD		
(lb)	2.1/3.1	2.7/3.9
(N)	9.3/13.8	12.0/17.4
Thickness nominal		
3/4-in. dia. area at 0.16 psi (mils)	16	21
19.05-mm dia. area at 1.10 kPa (mm)	0.406	0.533
Mullen burst		
(lb)	35	55
(N)	156	245
Frazier air permeability		
[(ft^3/min)/ft^2]	400	250
[(m^3/min)/m^2]	122	76
Water vapor permeability		
[(g/24 hr)/m^2]	900	1,000

Source: Ref. 11.

2.5.21 *Stabilenka*

Stabilenka™ is a brand name of the American Enka Company, Enka, North Carolina,[27] used to designate a family of nonwoven polyester fabrics developed specifically for soils engineering purposes. Generally, the product provides soil/water filtration and soil stabilization functions. Being a polyester product, Stabilenka filter fabric provides high resistance to outdoor deterioration and dimensional change. Three styles are available, with specifications given in Table 2.25.

2.5.22 *Supac*

Supac® is a registered trademark of the Philips Petroleum Company.[33] The fabric is a needle punched nonwoven polypropylene fabric made by the Philips Fibers Corporation. It is intended for use in separation,

TABLE 2.25. Properties of Stabilenka™ Filter Fabric

Properties	Style Number			Test Method
	T-80	T-100	T-140	
Fabric weight (oz/yd^2)	2.3	3.4	4.3	
Fabric thickness (in.)	0.02	0.03	0.03	
Roll width (in.)	42 and 84	42 and 84	42 and 84	
Roll length (linear yd)	547	547	547	
Tensiles dry grab strength (lb)				ASTM D-1682
Length	64	80	129	
Width	54	65	96	
Tensile dry grab elongation (%)				ASTM D-1682
Length	55	41	42	
Width	68	53	46	
Trapezoid tear (lb)	29	29	30	ASTM D-2263
Flow rate [(gal/min)/ft^2] [a]				
At ΔH = 30–10 cm	400	556	257	
At ΔH = 50–30 cm	758	956	630	

Source: Ref. 27.

[a] Flow rates were determined by measuring the time required for the height of column of water standing above the test fabric to drop from 50 to 30 cm and from 30 to 10 cm as the water passes through the fabric. The flow rate was then calculated in gallons per minute per square foot of fabric.

reinforcement, drainage, erosion control, and other construction applications. Table 2.26 gives some of its engineering properties.

2.5.23 Terrafix

Terrafix® filter mats were developed in the mid-1960s by Terrafix Erosion Control Products, Inc.[37] to provide erosion protection for hydraulic structures. The mats are composed of a labyrinth of individual synthetic fibers in a three-dimensional structure. Certain mat types are reinforced with a polyester scrim and synthetic binder. Stated goals of Terrafix filter mats are

- Retention of fine-grained soil particles while maintaining high filter permeability.

TABLE 2.26. Typical Properties of Supac® Soil Filter Fabric Type 5-P

Property	Value
Nominal fabric weight, (ASTM D-2646) (oz/yd^2)	5.3
Fabric thickness, (ASTM D-1777) (mils)	50
Tensile Properties (ASTM D-1682)	
Ultimate strength, warp direction, wet (lb)	125
Ultimate strength, filling direction, wet (lb)	150
Elongation at break, wet (%)	80
Toughness (product of strength and elongation—averaged)	10,000
Ultimate strength after abrasion (Taber abrader, CSI-17 wheel, ASTM D-1175)	114
Stretch,	
Elongation; 27 lb., 3-in. width for 10 minutes. CFFA-19 (%)	19
Set (% of unrecovered stretch. CFFA-19)	4
Trapezoidal tear, ASTM D-2263 (lb)	73
Puncture strength, ASTM D-751 modified (lb)	100
Mullen burst, ASTM D-751 (psi)	300
Permeability	
Air permeability at 0.5-in. water head, ASTM D-737 [$(ft^3/min)/ft^2$]	230
Water permeability, coefficient of, C. of E. EM 1110-2-1906 (modified) (cm/sec)	5×10^{-2}
Equivalent opening size, C. of E. CW-02215 (modified)	Fabrics of different values available

Source: Ref. 33.

- Reinforcement and support of structures located on unstable foundation materials.

The fiber diameters and density of the mats can be tailored to meet a given condition and can also be reinforced with woven fabrics or produced with multilayer construction if necessary. This gives rise to a number of different styles, as shown in Table 2.27.

2.5.24 Terram and Filtram

Terram® is ICI Fibers' trade name[38] for its fabrics for the civil engineering industry: these range from lightweight, thermally bonded, nonwoven,

TABLE 2.27. Physical and Mechanical Properties of Terrafix

Type	Fiber	Thickness (mm)	Weight (g/m^2; oz/yd^2)	Grab Tensile Strength, Wet (lb)	Mullen Burst (psi)	EOS U.S. Seive Number	Water Permeability at $\Delta H = 35$ cm [$(ml/cm^2)/sec$]
300 NA	Polyester	3.5 (137 mils)	500; 14.7	200 @ 25%	360	50	56
500 NA	Polyester	4.5 (177 mils)	700; 20.6	200 @ 25%	445	70	45
400 NR	Polyamide	4.0	550; 16	120 @ 25%	300	120	10
270 R	Polyester	3.0	270; 8	150 @ 100%	300	100	40
470 R	Polyester	4.0	470; 11	200 @ 100%	400	120	30
1000 R	Polyester	6.0	1,000; 29	425 @ 110%	800	120	10
1600 R	Polyester	6.0 (236 mils)	1,600; 47.2	700 @ 125%	Beyond machine capacity	120	5
814 B	Polyester plus polypropylene	6.0 (236 mils)	800; 22.1	350 @ 120%	536	120	16
1002 NS	Polyamide plus polyester	11.0 (433 mils)	1,110; 32.5	—	—	—	75
370 RS	Polyester	4 (157 mills)	475; 14	90 @ 30%	390	120	35
Terra-track 2415	Polypropylene	0.45 (18 mills)	125; 3.7	105 @ 20%	280	50	6

Source: Ref. 37.

permeable materials designed for use in ground stabilization, drainage, reinforcement, and erosion control (for properties see Table 2.28) to special-purpose fabrics with either unidirectional or two-directional strength (between 5 and 80 tons per meter) for soil reinforcement.

In addition, ICI has developed a laminated product, marketed under the trade name Filtram®, which promotes the flow of water within its own structure. This integrated, lamina, flexible filter drain (in which Terram is used as the filter membrane) eliminates the need for graded

TABLE 2.28. Structural Characteristics and Mechanical Properties of Terram from ICI Fibers

	Terram Product					
Property	500	700	1,000	1,500	2,000	3,000
Structural Characteristics						
Fiber diameter (μm)	35	35	35	35	35	35
Thickness in terms of fiber diameters	11	16	20	23	25	28
Porosity (% at 250 kg/m^2)	81	80	79	73	72	70
Porosity (% at 2×10^4 kg/m^2)	75	74	74	68	67	65
Mechanical Properties[a]						
1 m plane strain tensile						
Maximum load (kN/m)	3.75	6.0	8.5	11.0	13.0	14.0
Extension at maximum load (%)	35	40	45	50	55	60
Extension at break (%)	40	45	50	55	60	75
Load at 5% extension (kN/m)	1.0	2.0	2.5	3.0	3.5	3.75
25 mm grab tensile						
Maximum load (N)	400	600	850	1,200	1,400	1,600
Extension at maximum load (%)	70	75	80	80	80	80
Extension at break (%)	80	85	90	90	90	90
Tear strength–wing (N)	110	190	250	310	360	400
Burst load (N/cm^2)	50	80	110	150	180	210

Source: Ref. 38.

[a] Figures are comparable to equivalent figures obtained by testing to DIN 53857 and DIN 53858.

filters, and is of particular interest in situations where conventional drains are difficult to install.

Terram and Filtram are trademarks of Imperial Chemical Industries, Ltd., for its products for the civil engineering industry.

2.5.25 *Typar*

Typar® is a spun-bonded polypropylene developed by the duPont Company[11] as primary backing for tufted rugs and carpets. It is currently used in geotechnical engineering in separation, reinforcement, drainage, erosion, and other uses. Typar is a uniform sheet of preferentially oriented continuous filaments of 100 percent isotatic polypropylene manufactured by an integrated process of fiber spinning and bonding. There are two basic styles (3401 and 3601), as well as an EVA-coated impermeable style (T063). Table 2.29 presents the pertinent properties.

TABLE 2.29. Properties of Typar® Fabric

	Style Number			
Property	3401	3601	T063	ASTM Test
Weight (oz/yd^2)	4.0	6.0	7.5	ASTM D1910
Thickness (mils)	15	19	15.5	ASTM D1777
Grab tensile (lb)	130	225	180	ASTM D1682
Elongation to break (%)	62	63	68	ASTM D1682
Trapezoidal tear (lb)	70	75	54	ASTM D2263
Mullen burst (psi)	170	263	265	ASTM D774-46
Specific gravity	0.95	—	0.96	—
Equivalent opening size	70 to 100	140 to 170	—	CE/ASTM D422
Coefficient of H_2O permeability K (cm/sec)	2×10^{-2}	—	—	EURM-100
Modulus (lb)	1200	—	1150	ASTM D1682

Source: Ref. 11.

2.5.26 *Tyvek*

Tyvek® is a duPont product[11] made from 100 percent density polyethylene fibers by an integrated spinning and bonding process. The sheet is formed by spinning very fine polyethylene fibers and then bonding

TABLE 2.30. Physical and Mechanical Property Comparison of Fabrics

						Physical Properties			Mechanical Properties			
Section Number	Fabric	Style	Manufacturer or Agent	Fiber Type[a]	Process Type[b]	Weight[c] (oz/yd^2)	Thickness[d] (mils)	EOS[e] (Sieve Number)	Grab Strength[f] (lb)	Elongation[g] (%)	Burst[h] (psi)	Trapezoidal Tear[i] (lb)
2.5.1	Adva-Felt	TS200	Advance	1	1	6.0	80	90	115	89		
		TS300				8.0	130	70	225	101		
		TS400				10.5	170	60	300	110		
2.5.2	Bay Mills	154	Bay Mills	4	4	9.7	14					
		144				6.0	9					
		610				2.5	4					
		196				32	45					
2.5.3	Bidim	C22	Monsanto	2	1, 2	4.5	60	50*	115	85	225	62
		C28				6.0	75	50*	160	80	360	93
		C34				8.0	90	70*	255	75	400	125
		C38				10.0	110	100*	300	65	500	170
		C42				16.2	190	100*	610	60	850	250
2.5.4	Cerex	—	Monsanto	3	1	0.3	2.3		8		17	3.4
						0.4	2.5		12		20	4.3
						0.5	3.2		16		24	4.5
						0.6	3.4		21		29	5.5
						0.7	3.7		27		33	6.3
						0.85	4.2		32		36	6.5
						1.0	4.8		41		40	8.0
						1.5	7.1		53		52	11.0
						2.0	8.7		70		65	14.0
2.5.5	Cordura	500	duPont	3	4	6.3	17		413		495	
		1000				9.9	27		567		680	
2.5.6	Enkamat	7010	American Enka	3	5		354			50		
		7020					710			50		

2.5.7	Fibretex	320	Crown	1	1, 2	9.4			125	130		
		420	Zellerbach			12.4			150	150		
		600				17.7			250	160		
		200				5.9			—	—		
		300				8.8			—	—		
		400				11.8			—	—		
2.5.8	Filter-X	—	Carthage	5	4	11.6		100	200	23	260*	70*
			Mills									
2.5.9	Laurel	A	Laurel	1	4	7.2	17	100	400	33	528	90
	Cloth	B	Plastics			6.3	22	40	280	40	520	—
2.5.10	Mirafi	140	Celanese	5	3	4.1	30	—	120*	130*	—	65
		500X		1	4	4.0	25	—	200*	—	325	—
		100X		1	4	—	—	40	120*	—	200	65
2.5.11	Monofelt	—	J. P. Stevens	1	5	5	—	90	130	72	230	
2.5.12	Monofilter	—	J. P. Stevens	1	4	7	20	40	347	—	532	
2.5.13	Nicolon	66339	U.S.	5	5		30	40	260	30	500	
		66373	Textures				30	35	240	32	500	
		66424					24	70	240	30	>600	
		66392					24	70	240	30	>600	
		66186					20	100	400	30	>600	
		66475					89	35	1250	18	>1500	
		HD20,000					30	—	—	9	—	
		LD1,000					—	—	200	—	>325	
		X					18	85*	>350	30	>500	
		HD40,000					—	—	—	—	—	
2.5.14	Permealiner	M-1195	Staff	1	4	7.2	—	85	400	34	510	92
		M-1105	Industries	1	4	6.5	—	30	275	28	520	110
		ISS-1		1	2	5.0	14	60	110	22	300	30
		ISS-2		1	2	7.5	18	120	160	12	400	30
2.5.15	Petromat	—	Philips	1	2	4.1			115	65		
2.5.16	Polyfelt	—	Advance	2	2	7.8	127		228	101		47
2.5.17	Poly-Filter	X	Carthage	1	4	7.2		70	380	23		
		GB	Mills			6.6		40	200	23		
2.5.18	ProPex	II	Amoco	1	4	6.5	—	30	275	28	520*	110
			Fabrics									

TABLE 2.30. (Continued)

						Physical Properties			Mechanical Properties			
Section Number	Fabric	Style	Manufacturer or Agent	Fiber Type[a]	Process Type[b]	Weight[c] (oz/yd^2)	Thickness[d] (mils)	EOS[e] (Sieve Number)	Grab Strength[f] (lb)	Elongation[g] (%)	Burst[h] (psi)	Trapezoidal Tear[i] (lb)
2.5.19	Reemay	2006	duPont	2	1	0.6	6		11		12	
		2011				0.7	7		14		31	
		2014				1.0	9		22		31	
		2016				1.3	10		31		46	
		2024				2.1	12		52		65	
		2033				2.9	16		70		96	
		2408				1.1	12		16		21	
		2416				1.5	14		22		28	
		2431				2.4	18		45		46	
		2441				2.9	20		56		58	
		2470				5.8	32		117		88	
2.5.20	Sontara	8000	duPont	2	5	1.2	16		25	40	35	
		8002				1.9	21		40	40	55	
2.5.21	Stabilenika	T-80	American Enka	2	5	2.3	20		64	55		29
		T-100				3.4	30		80	41		29
		T-140				4.3	30		129	42		30
2.5.22	Supac	5-P	Philips	1	2	5.3	50		125	80	300	73
2.5.23	Terrafix	300 NA	Erosion Control	5	5	14.7	137	50	200*	25	360	
		500 NA				20.6	177	70	200*	25	445	
		400 NR				16	160	120	120*		300	
		270R				8	118	100	150*		300	
		470R				11	160	120	200*		400	
		1000R				29	236	120	425*		800	
		1600R				47.2	236	120	700*		BMC	
		814B				22.1	236	120	350*		536	

		1002NS				32.5	433	—	—		—	
		370RS				14	157	120	90*		390	
		2415				3.7	18	50	105*		280	
2.5.24	Terram	500	ICI Fibres	5	3							
		700										
		1000										
		1500										
		2000										
		3000										
2.5.25	Typar	3401	duPont	1	1	4.0	15	85*	130	62	170	70
		3601		1	1	6.0	19	155*	225	63	263	75
		T063		5	5	7.5	15.5	—	180	68	265	54
2.5.26	Tyvek	—	duPont	5	1							

[a] Fiber type:
1. Polypropylene
2. Polyester
3. Nylon
4. Glass
5. Other or combined

[b] Process type:
1. Nonwoven, spun bonded
2. Nonwoven, needled, or mechanically bonded
3. Nonwoven, thermal, or melt bonded
4. Woven
5. Other, combined, or not known

[c] ASTM D 1910.
[d] ASTM D 1777.
[e] Corps of Engineers CW-02215.
[f] ASTM D 1682 (machine direction with dry fabric).
[g] ASTM D 1682 (machine direction with dry fabric).
[h] ASTM D 774, D 231, D 751.
[i] ASTM D 2263 (machine direction).
*Signifies a different test than above or test modified in company literature. Many values have been averaged when a test range was given.

them together with heat and pressure. The resulting paperlike product comes in different weights and is used in packaging, signs, labels, maps, envelopes, and so on, but has not yet been used in construction engineering applications.

2.5.27 Summary of Fabric Properties

In summarizing Section 2.5, it is readily seen that there is little consistency of reported test data. This is most unfortunate, as cross referencing of one fabric to another would be very desirable. However, an attempt has been made to point out the obvious gaps that are in the available literature. Table 2.30 presents test details of the 26 fabrics listed in Table 2.4 along with their processing methods and selected physical and mechanical properties. It appears that the physical properties most often reported are weight, thickness, and equivalent opening size (EOS). These values are listed where available. Regarding mechanical properties, grab strength, elongation, burst strength, and trapezoidal tear results are available for many fabrics; hence they are also listed.

Considerable caution should be used in comparing the fabrics even on the information presented in Table 2.30 since test variations in the ASTM test were often noted in the commercial literature, or the test designation was not specifically referenced. Furthermore, as previously noted, the fabric properties may be changed by the manufacturer to meet particular situations and/or use conditions.

2.6 References

1. E. R. Kaswell, *Handbook of Industrial Textiles*, Wellington Sears, New York, 1963.
2. INDA, Association of the Nonwoven Fabrics Industry, 10 E. 40 St., New York, N.Y. 10016.
3. A. McGown, "The Properties and Uses of Permeable Fabric Membranes," Residential Workshop on Materials and Methods for Low Cost Roads, Rail and Reclamation Works, Leura, Australia, Sept. 1976, pp. 663–709.
4. ______, *Sample Specifications for Engineering Fabrics*, FHWA-TS-78-211, U.S. Dept. of Transportation, Federal Highway Administration, Washington, D.C., 1978.
5. Federal Highway Administration Grant No. DOT-FH-11-9353 to J. R. Bell and A. B. Hicks, Dept. of Civil Engineering, Oregon State University, Corvallis, Oregon, "Evaluation of Test Methods and Use Criteria for Filter Fabrics."
6. M. Lewis, "Bibliography on Construction Fabrics," Dept. of Civil Engineering, Oregon State University, Corvallis, Oregon, 1978.

7. *Proceedings of International Conference on the Use of Fabrics in Geotechnics*, April 1977, Vols. I, II, III, Assoc. Amicale des Ingenieurs, Anciens eleves de l'E.N.P.C., 28 rue des Saints-Peres, 75007 Paris.
8. C. R. Sissons, "Strength Testing of Fabrics for Use in Civil Engineering," *C. R. Coll. Int. Sols Text.*, 1977, pp. 287-292.
9. J. A. Finnigan, "The Creep Behavior of High Tenacity Yarns and Fabrics used in Civil Engineering," *C. R. Coll. Int. Sols Text.*, 1977, Vol. II, pp. 305-309.
10. Product Literature, Monsanto Textiles Co., Nonwovens Business Group G4WC, 800 N. Linbergh Blvd., St. Louis, Mo. 63155.
11. E. I. duPont de Nemours and Co., Inc., Textile Fibers Department, Wilmington, Del., "Transverse Elasticity Test (Preliminary)," EVRM 110, Oct. 1977, and other product literature.
12. S. L. Alfheim and A. Sorlie, "Testing and Classification of Fabrics for Application in Road Construction," *C. R. Coll. Int. Sols Text.*, 1977, Vol. II, pp. 333-338.
13. T. W. Lambe, *Soil Testing for Engineers*, Wiley, New York, 1951.
14. Product Literature, Mirafi 140, 100X and 500X, Celanese Fibers Marketing Company, 1221 Ave. of the Americas, New York, N.Y. 10036.
15. A. L. Rollin, Ecole Polytechnique de Montreal, personal communication.
16. Department of the Army, Corps of Engineers, Civil Works Construction Guide Specification CW 02215, Nov. 1977.
17. New York State Filter Fabric Flow Capacity Test, N.Y. DOT, Albany, N.Y., Oct. 1977.
18. R. M. Parks, "Engineering Fabrics, Their Uses, Evaluation Methods and Specifications," in *Highway Focus*, May 1977, pp. 63-71.
19. I. Oncu, "Investigation of Permeability and Filtration Characteristics of Non-Woven Fabric Filters," Internal Report to R. M. Koerner, Drexel University, Philadelphia, Pa.
20. R. G. Carroll Jr., Celanese Fibers Marketing Company, Charlotte, N.C., personal communication.
21. New York State Department of Transportation, Soil Mechanics Bureau, Filter Fabric Soil Retention Test, (Interim Procedure), Albany, N.Y.
22. "Comparative Chemical Resistance of Fibers," Bulletin X-48, E. I. duPont de Nemours, and Co., Wilmington, Del., March 1956.
23. "Light and Weather Resistance of Fibers," Bulletin X-203, E. I. duPont de Nemours and Co., Wilmington, Del., April 1966.
24. "Burial Deterioration of Geotechnical Fabrics," Div. of Chemistry, Natural Research Council of Canada, Ottawa, Canada.
25. Product Literature, Advance Construction Specialties Co., P. O. Box 17212, Memphis, Tenn. 38117.
26. Product Literature, Bay Mills Midland, Ltd., Midland, Ontario, Canada L4R 4G1.
27. Product Literature, America Enka Co., Subsidiary of Akzona Inc., Enka, N.C. 28728.
28. Product Literature, Crown Zellerbach, Nonwoven Fabrics Div., Camas, Wash. 98607.

29. Product Literature, Carthage Mills, Erosion Control Div., 124 W. 66th Street, Cincinnati, Ohio 45216.
30. Product Literature, J. P. Stevens Co., Inc., Nonwoven Products Div., Stevens Tower, 1185 Ave. of the Americas, New York, N.Y. 10036.
31. Product Literature, Nicolon Corporation, U.S. Textures Sales Corp., 4229 Jeffrey Drive, Baton Rouge, La. 70816.
32. Product Literature, Staff Industries, Inc., 78 Dryden Road, Upper Montclair, N.J. 07043.
33. Product Literature, Philips Fibers Corp., P. O. Box 66, Greenville, S.C. 20602.
34. Product Literature, "ProPex," Amoco Fabrics Company, Patchogue Plymouth Division, 555 Interstate North, Atlanta, Ga. 30339.
35. Product Literature, Allied Chemical Corp. Fabric, Distributed by Erosion Control, Inc., 1655 Palm Beach Lakes Blvd., West Palm Beach, Fla. 33401.
36. *J. Coated Fabrics*, Vol. 6, April, 1977, pp. 208–209.
37. Product Literature, Erosion Control Products, Inc., 1329 Martin-grove Rd., Rexdale, Ontario, M0W4X5, Canada.
38. Product Literature, ICI Fibers, Melded Products Business Area, Hookstone Road, Harrogate, HG 2 80 N, Yorkshire, England.

3

Fabric Use in Separation of Materials

The separation of one material from another is often:

- Necessary and convenient.
- Necessary and impractical.
- Expensive.
- Wasteful.

There are many situations where separation can be done with fabrics. The reasons for separating the materials are varied, and three classes of problems come to mind:

- Soil to soil.
- Soil to stone or aggregate.
- Asphalt to asphalt (described in Section 4.5).

There are obviously other material systems that could be separated, but the above serve to illustrate the unique use of fabrics in this regard.

This chapter illustrates the use of synthetic fabrics in two important areas of construction activity: one, its use in zoned earth and earth/rock dams; the other, its use for separation of railroad ballast or stone base from the soil subgrade. This latter case is typical of any such separation of stone base from soil subgrade, as in roads, parking lots, driveways, sidewalks, and so on. In such situations the combined effect of the stone aggregate penetrating the soil subgrade and the soil subgrade infiltrating the stone aggregate decreases the permeability of the aggregate to the point where it cannot adequately transport the water that comes to it. The entrapped water leads to loss of strength, settlement, frost action, potholes and rapid pavement deterioration. This general application represents a potentially huge market for synthetic fabrics.

3.1 Zoned Earth Dams

Background

The complexity of modern zoned earth and earth/rock dams can be seen[1] in a few "typical" examples, as shown in Figure 3.1. Such sections, along with the transition zones that are required between the different zones, cause the earthmoving contractor incredible labor and logistics problems. With different materials coming from different borrow pits, or being blended on site, the elimination of any zone or transition zone material is indeed welcomed. The design engineer, too, is interested in the elimination or separation of the zoned materials for both simplicity in construction and ultimately improved economy to the owner.

As a separating material the fabric could be permeable or impermeable, but with modern construction fabrics (either woven or nonwoven) permeability is usually a desirable option as long as adequate strength can be achieved.

Case History

The first reported use of construction fabrics in earth dams on a large scale was by Terzaghi and Lacroix[2] in connection with the history-making design and construction of the Mission Dam in British Columbia, Canada from 1957 to 1960. Located in the valley of the Bridge River, a 60-foot-high diversion dam had been built in 1948 for a small power station. The installation of a larger power plant required a 200-foot-high dam to be placed at the same site without interfering with the operation of the existing plant. A cross section of the valley site and of the two dams is shown in Figures 3.2 and 3.3, where two pervious acquifers, separated by a highly compressible clay stratum, can be seen. Differential settlements of up to 15 feet were anticipated, which led to a zoned dam requiring major interaction between the different materials. The upstream slope consisted mainly of a stony till covered with a clay blanket for seepage control in the embankment area. The problem, however, was felt to be a stretching of the clay blanket between areas high on the slope, which settle very little, and the zone of a large settlement at the bottom of the slope. In addition, narrow zones immediately downstream from the sheet pile cutoff wall were also considered vulnerable. Therefore, all of these zones were covered with a plastic membrane to control cracking. This area is shown cross hatched in Figure 3.4. The fabric used was a continuous polyvinyl chloride plastic sheet, which was considered to be a "watertight, stretchable membrane." The reasoning behind its use, as expressed

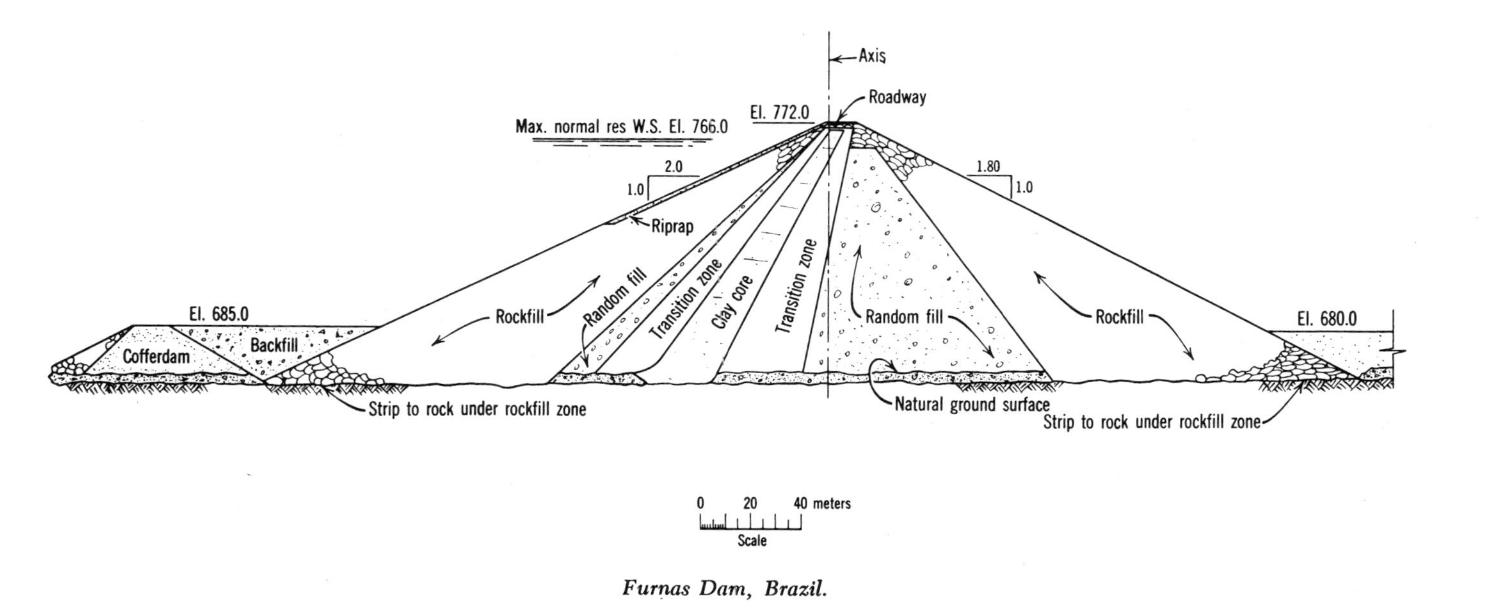

Furnas Dam, Brazil.

FIGURE 3.1. Typical designs of zoned earth and earth/rock dams (from Ref. 1).

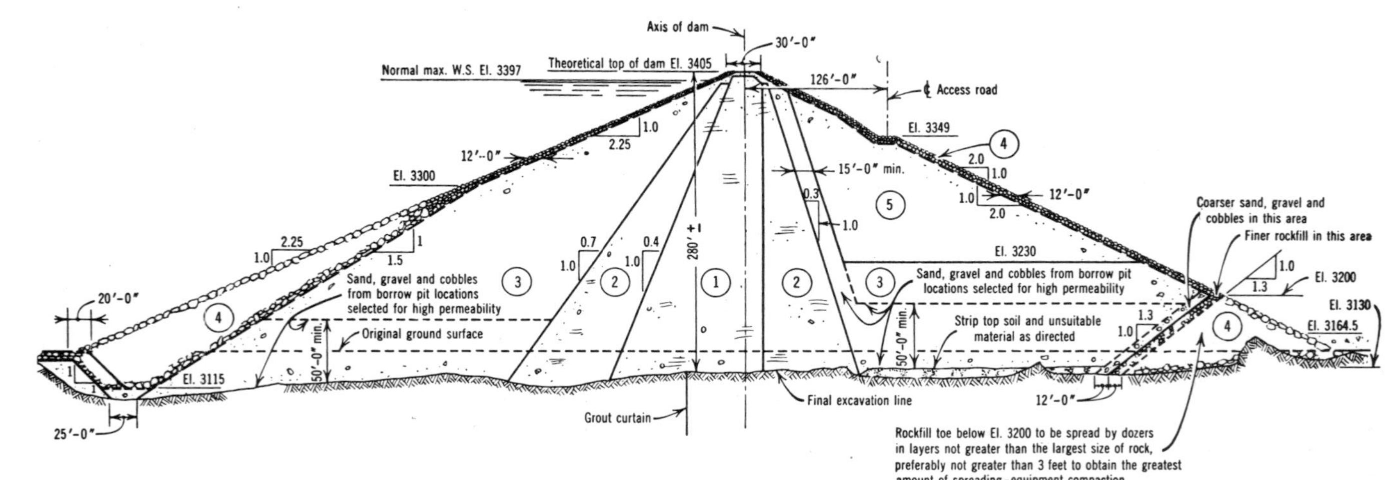

(1) *Impervious rolled earth—to consist of a mixture of silt, fine to medium sand, and gravel not exceeding 3 in. max. dimension.*
(2) *Semipervious rolled earth, transition zone—to consist of fine to coarse sand and gravel not exceeding 3 in. max. dimension, together with some silt sizes. The finest material to be adjacent to zone* (1) *with a gradual transition outwards.*
(3) *Rolled sand, gravel, and cobbles—the finest material to be adjacent to zone* (2) *and the coarsest adjacent to zone* (4) *and/or* (5).
(4) *Dumped rockfill—to consist of rock fragments from required rock excavations and primarily composed of well-graded fragments larger than ¼ ft.³ in volume with only enough rock spalls and gravel to fill voids in the coarser material.*
(5) *Random fill—to have a min. dry weight = 115 lb./ft.³ and min. angle of internal friction = 35°*

Beardsley Dam, California.

FIGURE 3.1. *Continued.*

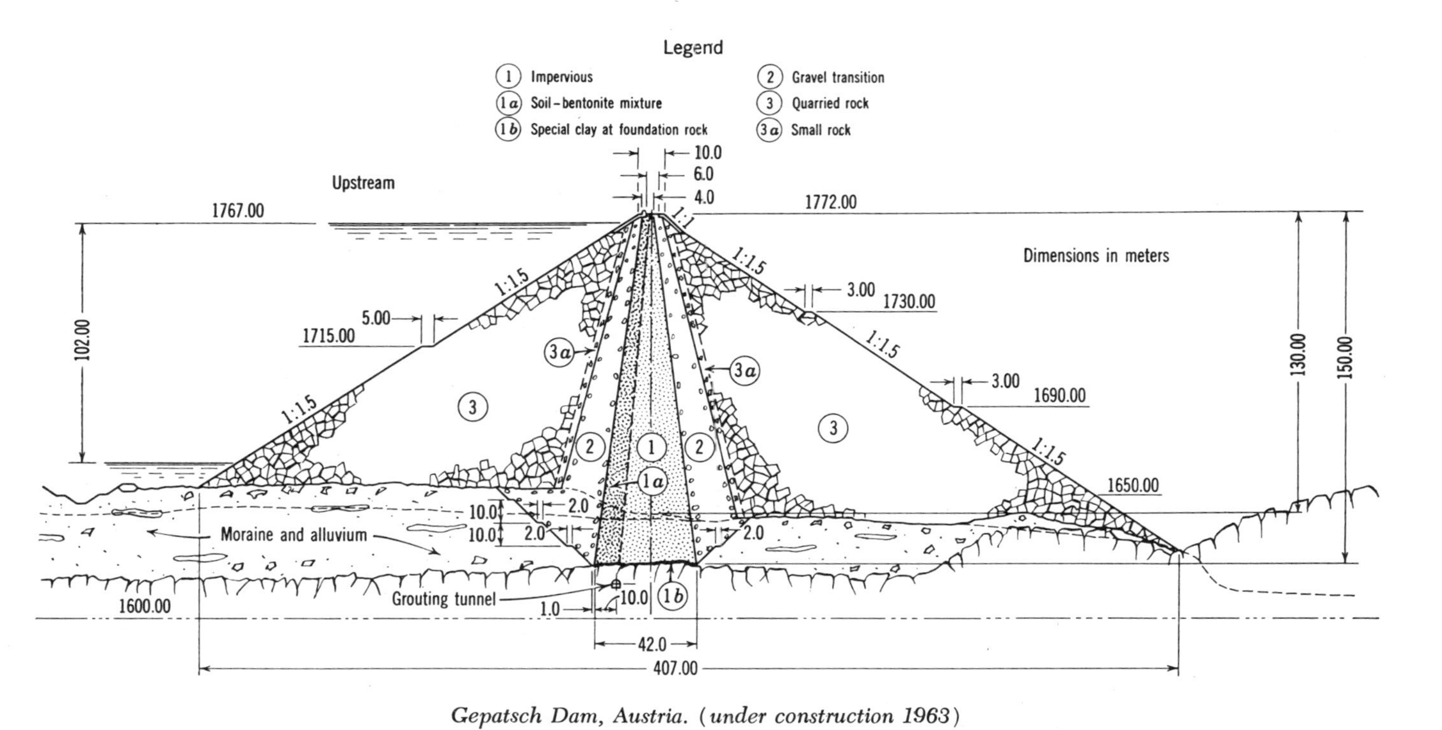

Gepatsch Dam, Austria. (under construction 1963)

FIGURE 3.1. *Continued.*

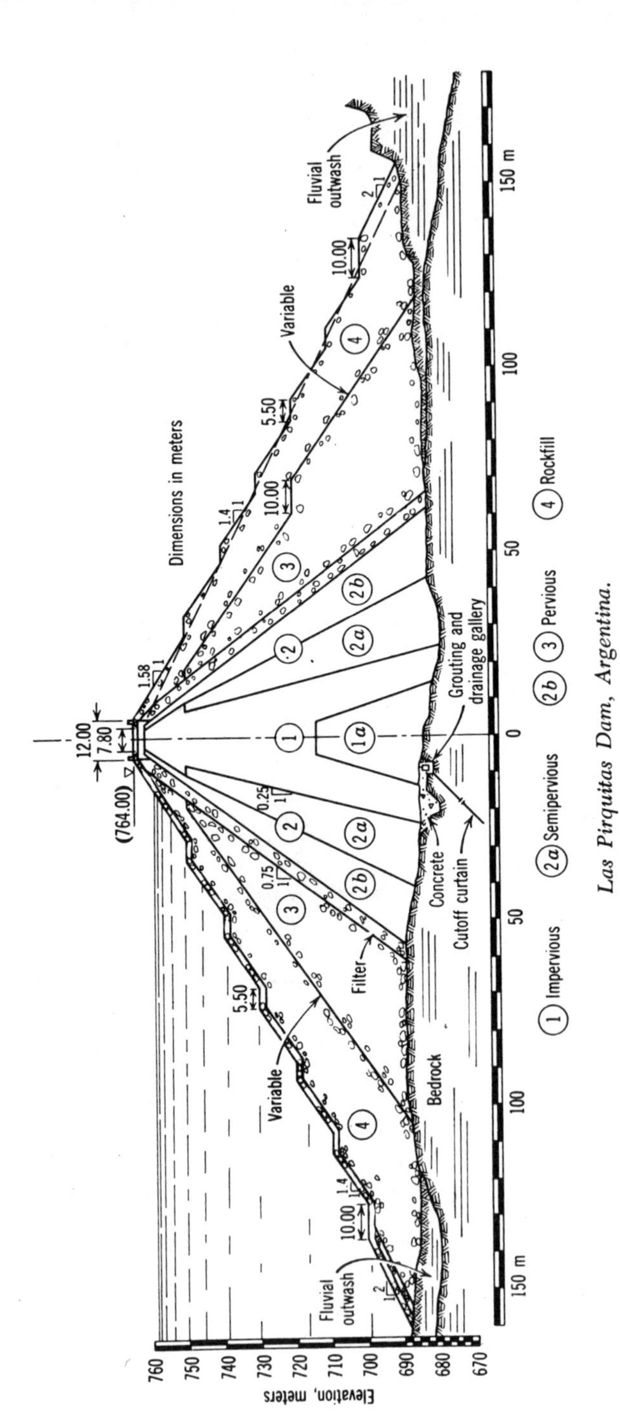

Las Pirquitas Dam, Argentina.

FIGURE 3.1. *Continued.*

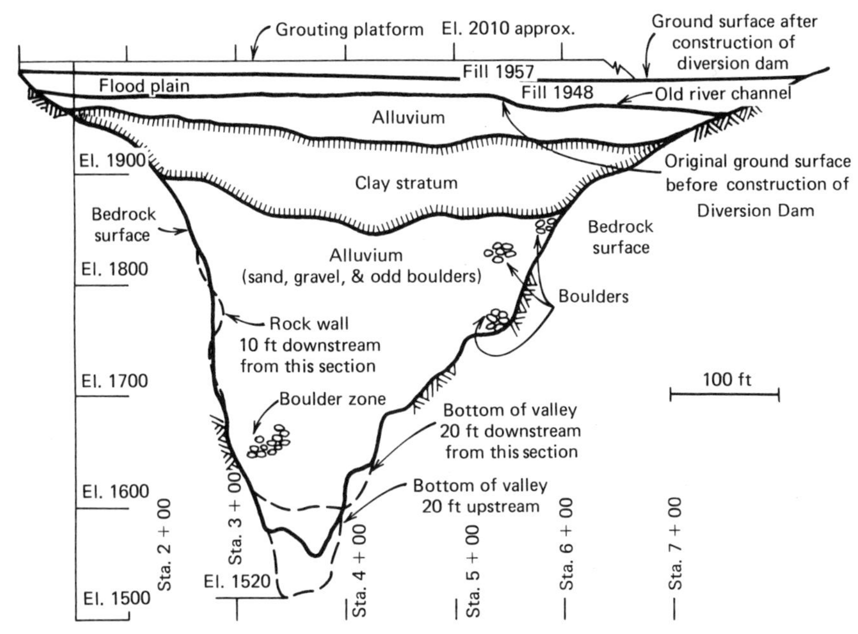

FIGURE 3.2. Cross section at site of grout curtain, looking downstream (from Ref. 2).

by Terzaghi and Lacroix, was that as soon as the overburden pressure acting in the membrane (protective fill plus reservoir load) exceeds the unconfined compressive strength of the clay, the resulting crack closes up and the integrity of the blanket is reestablished. Due to the expense of installation and the care required in placement, the membrane was placed only in zones of maximum tensile strains.

This breakthrough use of fabrics (albeit not of the type of modern construction fabrics available currently) was backed up by an extended series of laboratory tests. These tests, reported by Lacroix in 1960,[2] consisted of testing the strength of various types of shop joints and field joints in splicing the membrane and of three different types of tests to measure the adhesion between the membrane and the soil. These latter tests consisted of a modified direct shear test, a weighted sled on a plane of variable inclination, and a sled pulled in a horizontal plane.

While this early use of a plastic membrane was remarkable in itself, the same project saw another innovative use of fabrics. Completely described in the construction of the seepage cutoffs through the acquifers was the use of fabrics as forms, which is treated in Chapter 7.

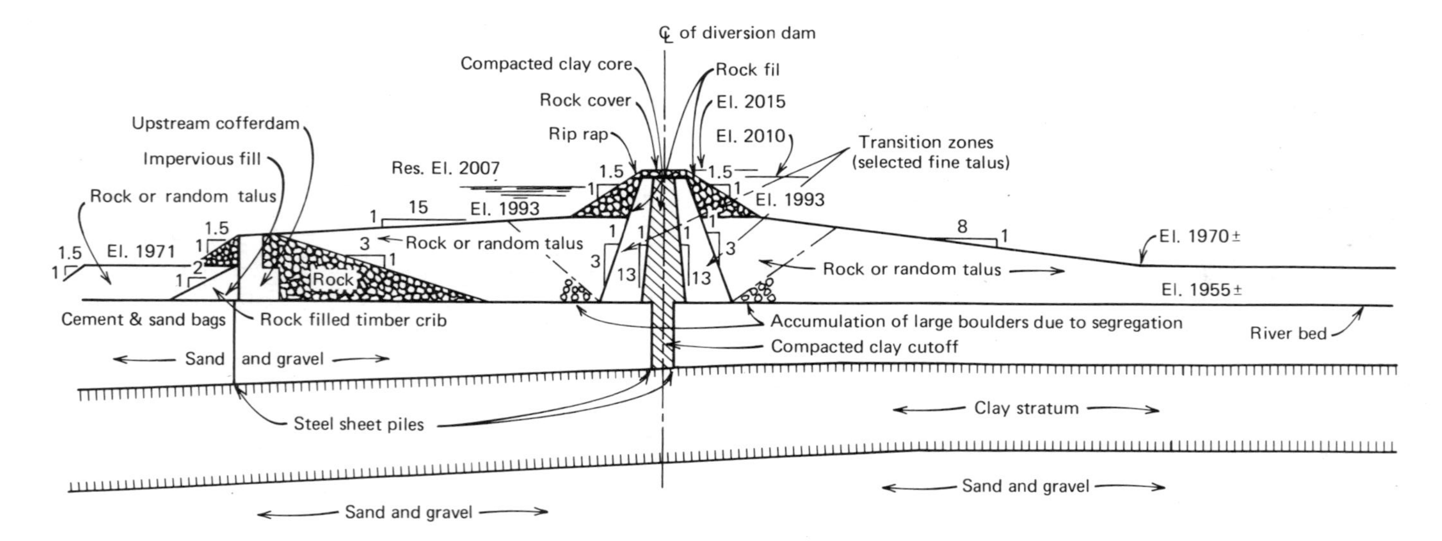

FIGURE 3.3. Cross section of Diversion Dam (from Ref. 2).

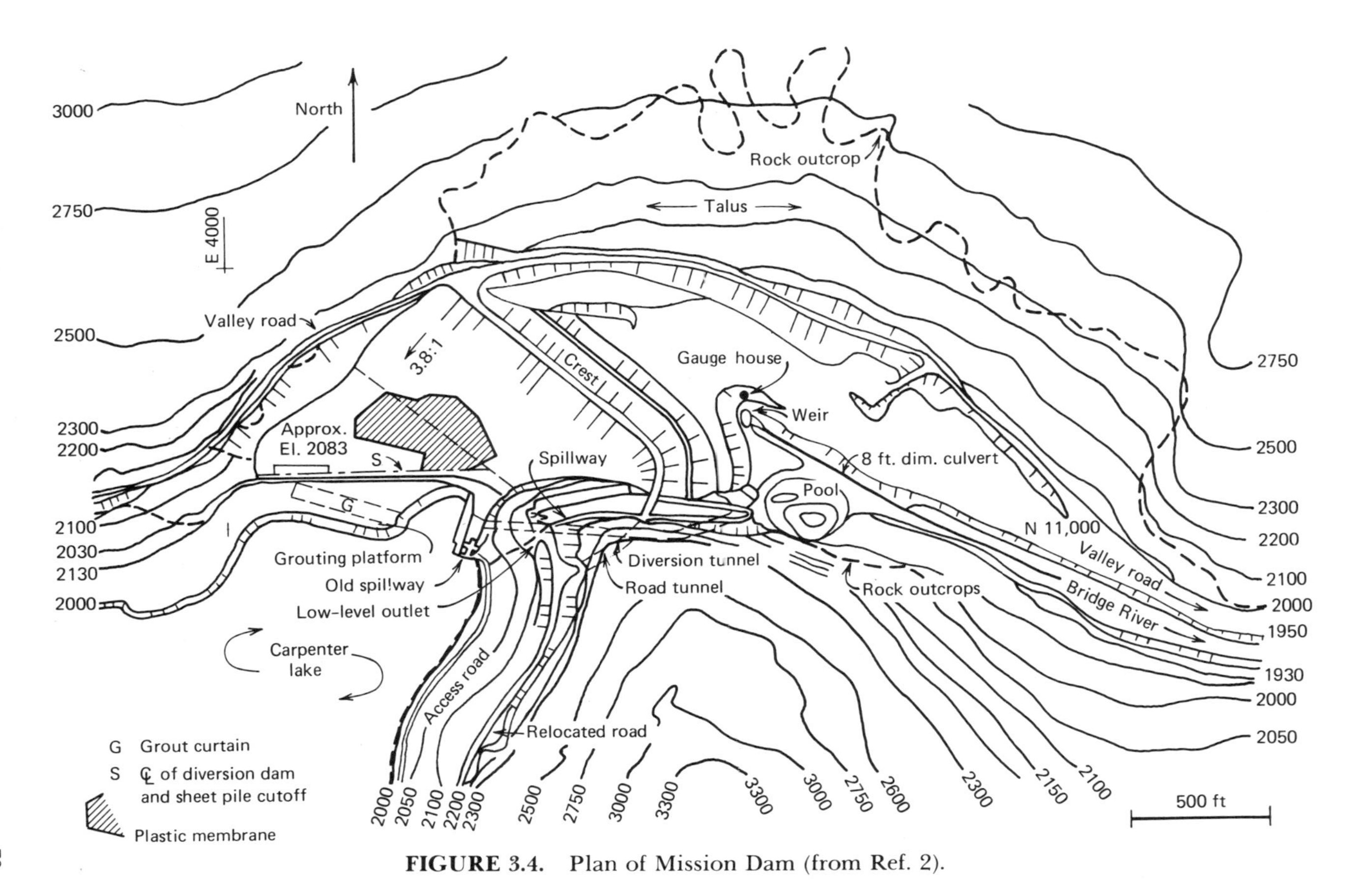

FIGURE 3.4. Plan of Mission Dam (from Ref. 2).

3.2 Railroad Ballast/Subgrade Separation

Background

Typical cross sections of railroad beds consist of a compacted and graded subgrade, 6 to 24 inches of compacted stone base, 6 to 18 inches of stone ballast, ties, and track. The stone base provides a number of valuable functions, for example, providing support when weak subgrades are encountered (reinforcement), prohibiting the soil subgrade from entering into the voids of the ballast and of the ballast from punching into the soil (separation), and allowing for free passage of water (drainage).

The elimination of all or part of this stone base has become a prime target for fabric utilization on the basis of lower material cost, lower placement cost, and quicker installation time. Additionally, the proper method of placing a stone base course by first placing the large stones in a compacted state, then screening in fines to fill the voids is almost passé. Current practice is to place the entire well-graded granular mass in one lift. For a dynamic load such as a railroad this practice seems questionable, since the high vibration levels will cause a readjustment of the large-size stones, causing both settlement and lateral shifting of the base course.

Case Histories

CASE 1. Since 1973 the German Federal Railway has been field testing fabrics for the purpose of;

- Eliminating the 20 to 30 centimeters of graded gravelly sand filter layer beneath the ballast (this has the function of preventing migration of the subgrade soil into the ballast).
- Increasing the bearing values of loads placed on top of the ballast when fabric is incorporated.

The first application is therefore as a separation mechanism and the second as a reinforcement mechanism.

Einsenmann and Leykauf[3] present the results of large-scale laboratory tests in which the fabric Terram prevented upward movement of fines in a similar manner to a conventional graded filter layer. The fabric withstood the imposed loading, and considerable plastic deformation, without damage. The reinforcement effect was also noticed, especially where subgrades of low bearing capacity were used or where track loading was high.

CASE 2. In a series of four articles[4] in *Railway Track and Structures* magazine, description of fabric installation beneath existing track and intercrossing is discussed. Bidim, Mirafi 140, Typar, and Polyfelt TS 300 installations are each described but in very sketchy detail. The general functions that track men see in the utilization of fabrics are separation, drainage, tensile reinforcement, and planar flow. However, the dominant function is that of separation of the soil subbase from the gravel ballast. We feel that this feature alone will potentially eliminate all, or much of, a multi-million-dollar annual ballast cleaning program in the United States, allowing for faster and safer railroad service and less maintanence in the future.

Regarding construction where track cannot be removed, the fabric can be installed by using a Plasser ballast undercutter cleaner. This piece of railroad equipment utilizes a chain saw type of arm which swings beneath the track it is sitting on and removes the entire ballast layer, moving it to a trench dug on the side of the track. While moving forward, the machine automatically unrolls the fabric onto the exposed subgrade. The new, or previously removed and cleaned ballast is then placed on top of the fabric, tamped beneath the ties, and tracks are then leveled using conventional track equipment.

The following installations are mentioned in the series of articles[4] but no performance data is offered.

- Station platforms of Chicago and North Western Railroad.
- Road crossings in Texas.
- Test sections in the South and Southwest areas of the United States.
- Main line track on the Illinois Central Gulf near Paxton, Illinois.
- Realignment and grading in New York State.
- Spur track to plant in Seneca, Illinois by the Rock Island Railroad.
- Marshalling yard at a plant in Indiana.

CASE 3. An interesting application of fabrics for use at railroad refueling locations and track areas where spills are likely to occur is shown in Figure 3.5. Here a double fabric system is used for separation, containment, and removal of the spilled liquids.

The separation function is provided for by an upper layer of permeable fabric, which prevents sand and silt from fouling ballast while permitting spilled oil to pass through it. The containment and removal functions are provided for by means of an impermeable fabric beneath the ballast, which allows for gravity flow to a suitable collection and removal system. True Temper Corporation, which has suggested the

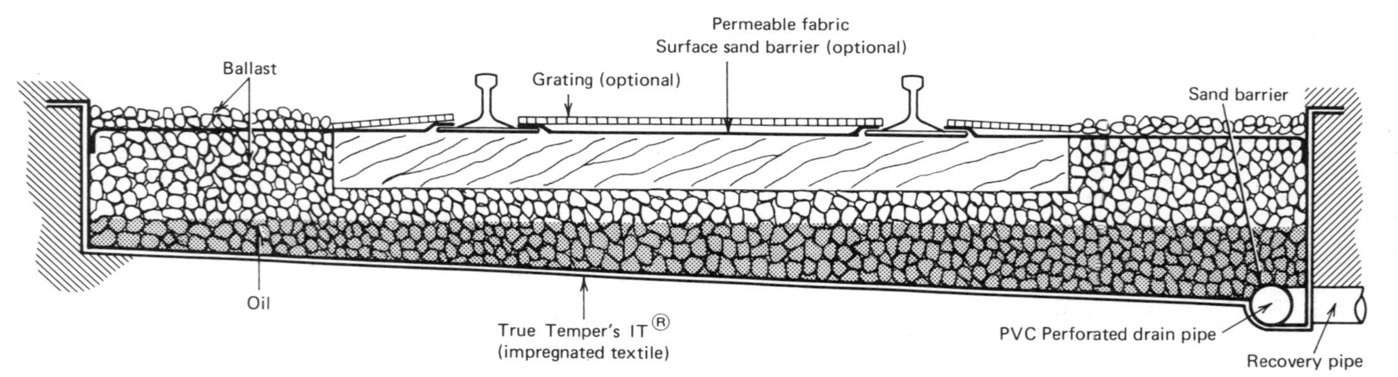

FIGURE 3.5. Suggested use of fabrics as an oil containment system at railroad refueling locations (after True Temper Corp.).

concept, recommends their IT® impregnated textile for the lower membrane and a needle punched polyester fabric for the upper membrane.

3.3 References

1. J. L. Sherard, R. J. Woodward, S. F. Gizienski, and W. A. Clevenger, *Earth and Earth-Rock Dams,* Wiley, New York, 1963.
2. K. Terzaghi and Y. Lacroix, "Mission Dam: An Earth and Rockfill Dam on a Highly Compressible Foundation," *Geotechnique,* March 1964, pp. 13-50.
3. J. Einsenmann and G. Leykauf, "Investigation of a Nonwoven Fabric Membrane in Railway Track Construction," *C. R. Coll. Inst. Sols. Text.*, 1977, Vol I, pp. 41-45 (in German).
4. ———, "Testing of Subgrade-Stabilization Fabrics Moves Ahead," *Railway Track and Structures,* Part 1, July 1976, pp. 20-21; Part 2, September, 1976, pp. 32-34; Part 3, October, 1976, pp. 22-23; and Part 4, December, 1976, pp. 18-19.

4

Fabric Use as Reinforcement

That very day, Pharaoh gave to the supervisors of the lower classes as well as to the scribes the following order: "Do not furnish these people with straw to make bricks, as you did previously. Henceforth let them go find the straw they need." (Exodus V:6–9). This quotation[1] certainly set the tone for the use of fabrics to strengthen materials. Obviously the intuitive feeling gained by our forefathers was in the right direction.

Today, in construction materials engineering, we usually think of the stress versus strain behavior of the material. For a fabric imbedded in soil, Broms[2] has conducted an interesting series of laboratory triaxial shear tests. His work, shown in Figure 4.1, is for dense and loose sands at low and high confining pressures. These results lead to the following conclusions.

- Failure stress is approximately doubled when the fabric is located properly within the test specimen (at the third points for this test configuration).
- Failure stress is not measurably affected when the fabric is improperly placed (as in the dead zones at the top and bottom of the sample).
- The material is stiffened (has a higher modulus of elasticity) when the fabric is properly placed.

Now that we know how fabric improves the performance of soils in the laboratory, it remains to see if this knowledge can be properly applied in the field.

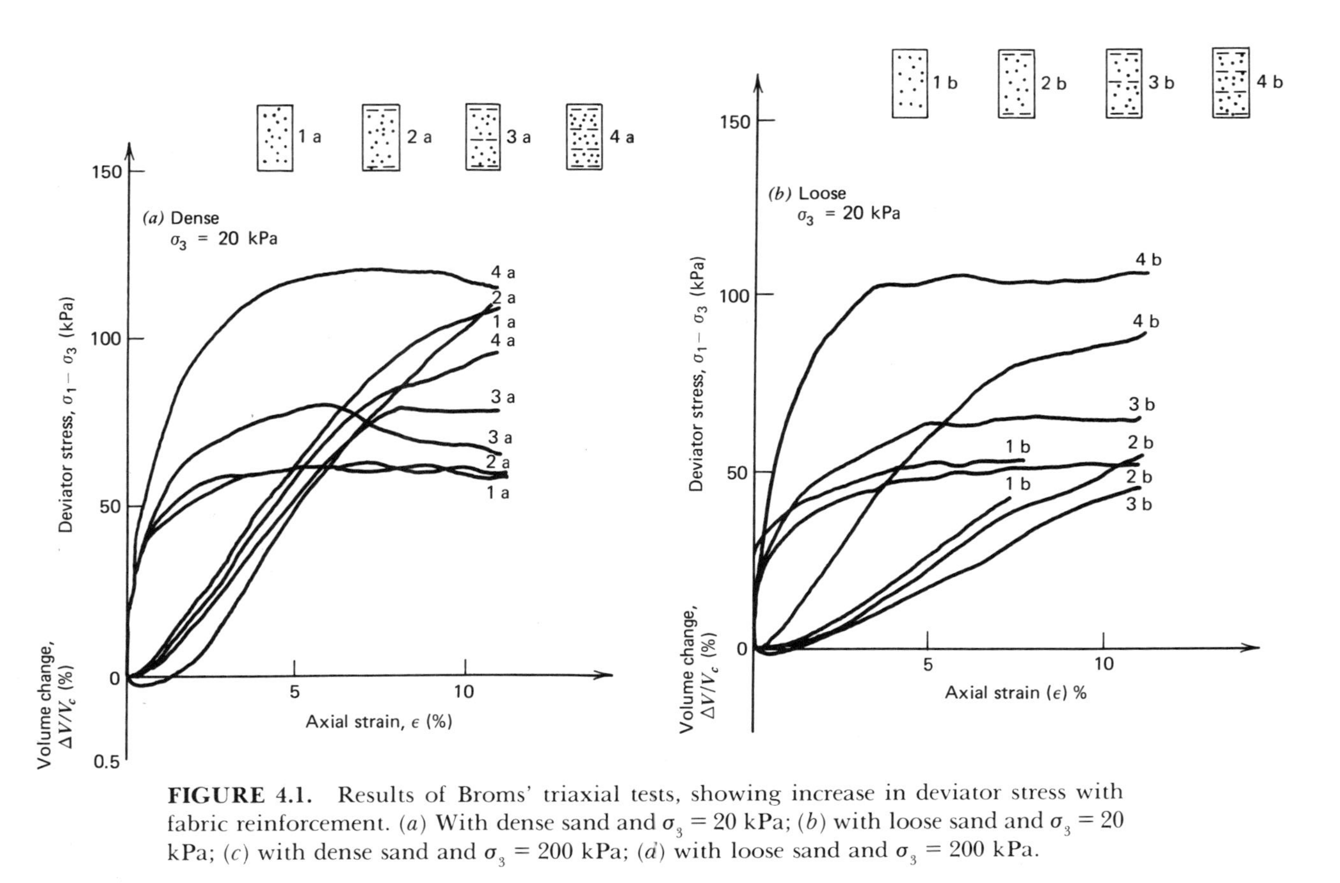

FIGURE 4.1. Results of Broms' triaxial tests, showing increase in deviator stress with fabric reinforcement. (*a*) With dense sand and $\sigma_3 = 20$ kPa; (*b*) with loose sand and $\sigma_3 = 20$ kPa; (*c*) with dense sand and $\sigma_3 = 200$ kPa; (*d*) with loose sand and $\sigma_3 = 200$ kPa.

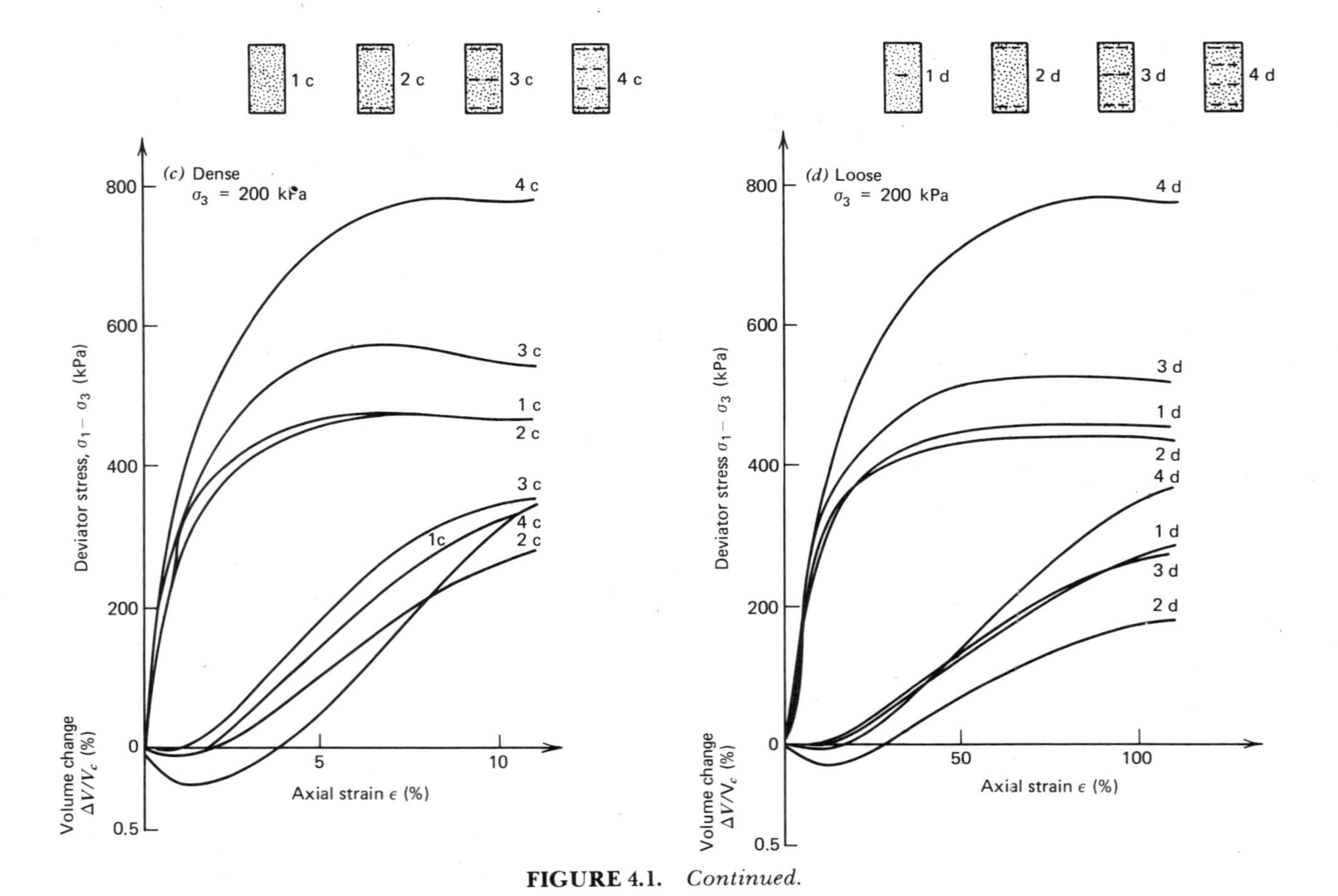

FIGURE 4.1. *Continued.*

4.1 Fabrics in Road Construction

Introduction

In describing the role of fabrics in reinforcing weak soil subgrades for road construction it is relatively common to refer to the Boussinesq Theory of stress mobilization in an elastic, semi-infinite half space. In so doing vertical stresses[3] as calculated by the following equation,

$$\sigma_z = \frac{P}{2\pi z^2}(3 \cos^5 \theta)$$

where σ_z = vertical stress
P = applied surface load
z = depth to point considered
θ = angle from load application to point considered

are generated and shown in Figure 4.2(*a*), before fabric placement, and approximated after fabric placement, as shown in Figure 4.2(*b*). Although not quite technically accurate, it is easily seen that stress intensity beneath

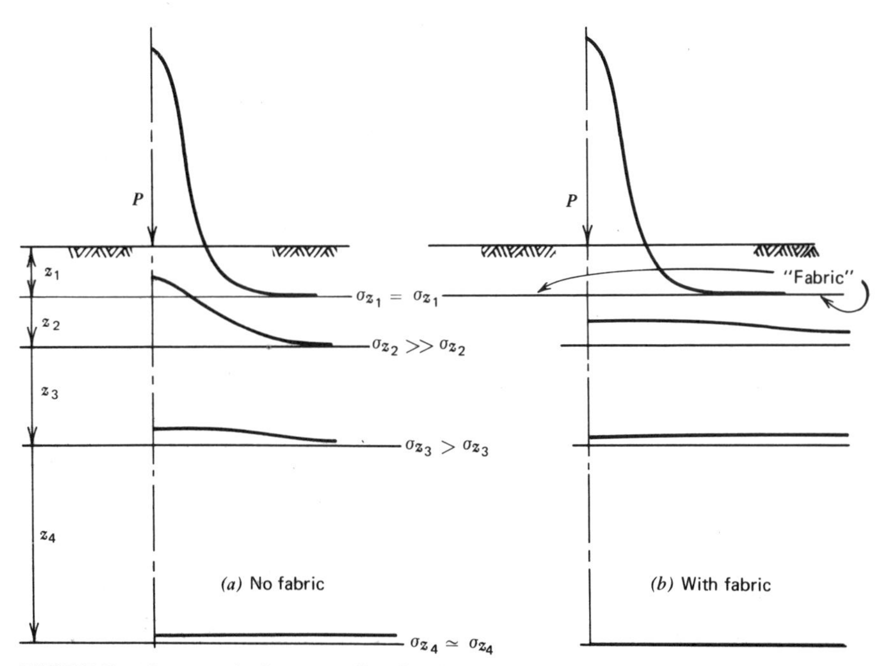

FIGURE 4.2. Vertical stress distribution without fabric (Boussinesq theory) and with fabric (approximate).

the fabric is reduced by its placement. Soil above the fabric is then replaced by a suitable aggregate to make the completed section.

We feel that the role of fabrics in road construction can be better illustrated by using the radial stress distribution of Boussinesq,[3] as calculated from equation 4.2,

$$\sigma_r = \frac{P}{2\pi z^2}\left(3 \sin^2 \theta \cos^3 \theta - \frac{(1 - 2\mu) \cos^2 \theta}{1 + \cos \theta}\right) \qquad (4.2)$$

where σ_r is the radial stress; P, z, and θ are as defined above, and μ is Poisson's ratio of soil, which is plotted in Figure 4.3. Here it is seen that directly beneath the surface load the soil is in tension. This tensile stress is maximum near the surface and dissipates with depth. It also spreads out radially at approximately 17°, as shown in Figure 4.3. Using this concept, but now extending it to a uniformly distributed load, as shown in Figure 4.4, it is seen that the entire zone beneath the load is in tension. If fabric is placed within this zone it, too, will be in tension. As the test data of Figure 4.1 show, the interaction of properly placed fabric and soil

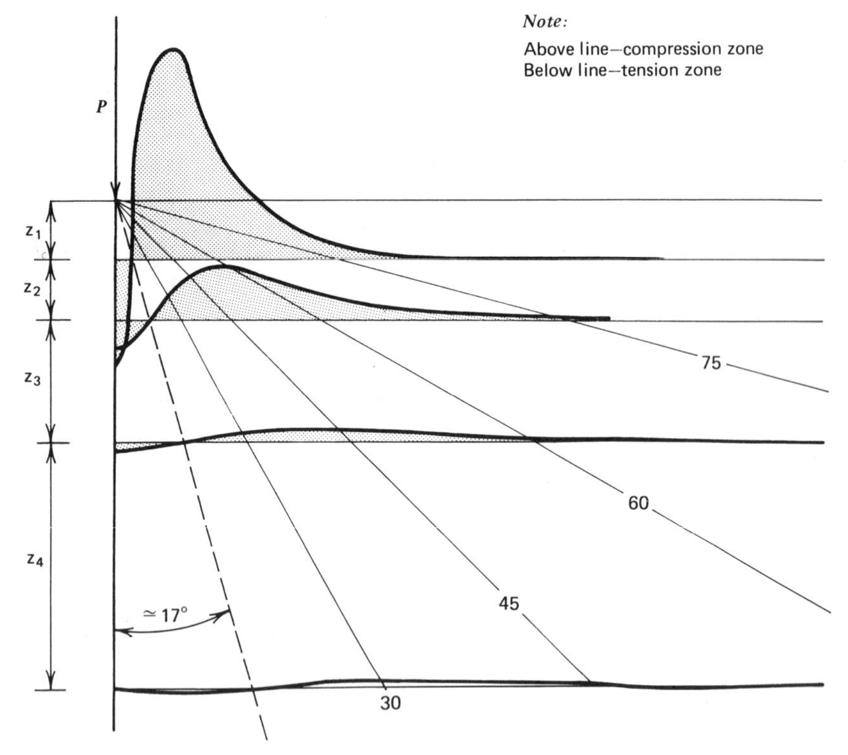

FIGURE 4.3. Radial stress distribution (Boussinesq theory).

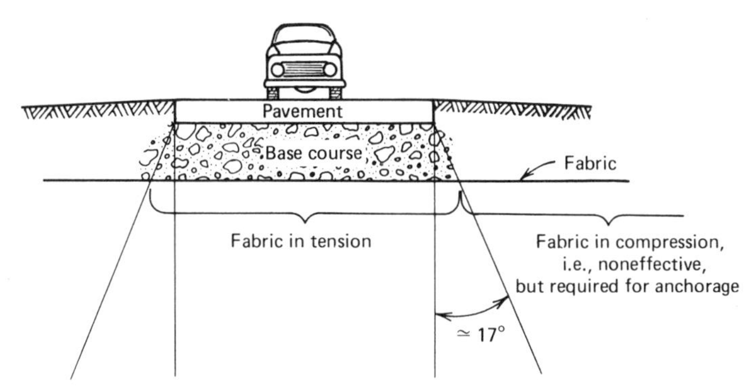

FIGURE 4.4. Practical adaptation of results of Boussinesq's radial stress distribution for a distributed surface load as occurs in a roadway section.

can be put to good advantage in increasing the load-carrying capacity of the soil in its unreinforced condition.

Background

The first use of fabrics (of the type of construction fabrics under consideration in this book) to reinforce roads was begun in 1926 by the South Carolina Highway Department. They used cotton fabric on a previously primed earth base, applied hot asphalt over the cotton fabric, and then covered the asphalt with about 50 pounds of sand per square yard. By 1935 their work has progressed to the point that a paper describing eight separate field experiments was published.[4] The cotton fabric used was classified as a cider duck and weighed about 7 ounces per yard of 40-inch-wide material. Adjacent sections of fabric were sewn together. Some of the experiments, which lasted for nine years, showed the fabric to be in good condition and the road likewise. The general conclusion was that the fabric reduced cracking, raveling, and failures.

In the many years since these early tests, highway traffic has vastly changed in character and so have highway pavements. Today a typical road cross section consists of a graded and compacted subgrade (i.e., the *in situ* soil) covered with crushed stone aggregate base course, followed by an appropriate wearing surface. Over weak subgrades a subbase between the stone base course and the soil subgrade is commonly found to be economical. Three important factors play a role in the design of such a system.

1. Suitability of subgrade soil as identified by its CBR value or plate-bearing value.

2. Thickness of stone base aggregate as determined by the subgrade soil and the type and volume of traffic to be carried on the road.
3. Type of wearing surface as determined by the type and volume of traffic as well as environmental conditions.

When fabrics are placed between the soil subgrade and crushed stone base course, two vital functions are served.

1. The fabric acts as a separator between the soil and the stone, preventing an intermixing of the two materials. This prevents contamination of the stone by the underlying soil, thereby maintaining its required high permeability and transmissibility characteristics as well as the "loss" of the stone (with its high strength characteristics) into the weaker soil subgrade, that is, intrusion. As such, this topic could easily be placed in the previous chapter, where fabric use as separators is discussed.
2. When soil subgrades are poor, that is, consist of soft, compressible soils, the fabric plays a major role of reinforcement. Thus the topic is included in this section. The following example shows how fabrics can be of major benefit in this regard.

In preliminary laboratory tests, duPont[5] has determined that fabric placed on a soil will, acting alone, add as much as 4 CBR percent to the soil. (CBR refers to California bearing ratio and is a commonly used test method to determine the adequacy of stone base courses and soil subgrades. The higher numbers refer to strong materials and the lower numbers to weak materials.) Realizing that this is a very simplistic generalization, it can nevertheless serve to show the effectiveness of using fabrics as a reinforcement for poor soils. This effectiveness can be directly measured by a reduction of the required thickness of the stone base course.

The Asphalt Institute's method[6] of designing a flexible pavement system is based on the CBR of the soil subgrade as well as the traffic using the road. Their design charts (see Figure 4.5) allow for the determination of T_A, the total thickness using asphalt alone, and T_{min}, the minimum asphalt thickness. The difference $T_A - T_{min}$ multiplied by a suitable constant (2.0 is used in this example) gives the required thickness of stone base, T_{SB}. An example using a design traffic number of ten and varying the soil subgrade's CBR, results in the upper curve of Figure 4.6. Repeating the procedure for the same types of soil subgrades but now adding 4 CBR percent for the inclusion of fabrics between the soil and the stone base results in the lower curve of Figure 4.6. The difference in required thickness of stone base (the ordinate distance between the curves) represents the savings due to fabric reinforcement. It is seen that this reduc-

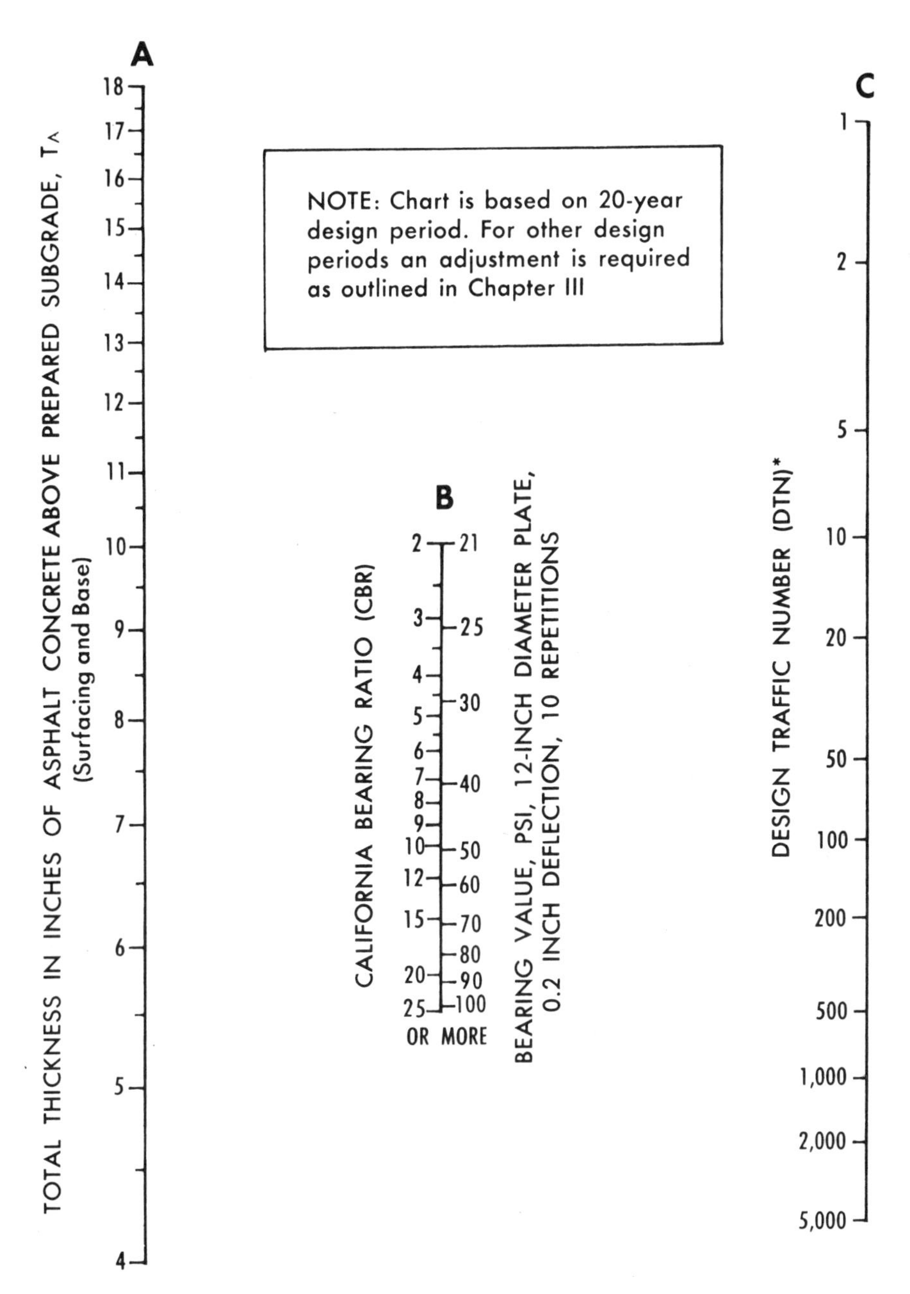

FIGURE 4.5. Typical design nomograph for determination of asphalt and stone base thicknesses for flexible highway pavements (after Asphalt Institute, Ref. 6).

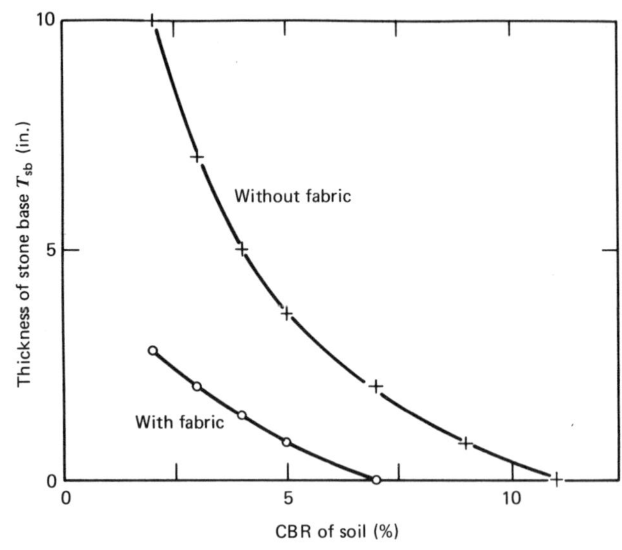

FIGURE 4.6. Typical results of required thickness of stone base as a function of CBR of soil.

tion in required stone base thickness becomes more pronounced with the low-CBR soils. Thus fabrics provide more reinforcement benefit with the softer soil subgrades than with firm soils of high CBR values. For the example shown a soil CBR of 2 results in savings of the required stone base thickness of $\frac{2}{3}$ (from 10 to 3 inches) when using fabrics. There is obviously a lower limit, as yet undetermined, at which this type of analysis cannot be extended. For example, water with a CBR of zero cannot be assumed to have an equivalent CBR of four when covered with fabric! The example does, however, show the general type of benefits accrued by the use of fabrics over weak soils.

What the example does not show is that fabrics also provide a very beneficial function of separation, that is, they prevent intrusion of soil subgrade into the stone base course, which is difficult to quantify. This separation function of the fabric provides its real benefit in time over the life of the pavement system and should not be overlooked.

Placement Procedure

The proper placement procedure for the use of fabrics in road subgrade reinforcement obviously depends on many factors, which make each job somewhat unique. Nevertheless, some generalized comments, common to most situations, can be made.

- The site should be cleared of debris, sharp objects, trees, and tree stumps (at least cut off at ground level) and similar objects that would cause the fabric to locally deform a large amount when fill is placed upon it.
- If construction machinery can move on the site, the site should be at least rough graded with holes being filled in and high spots being cut off. Compaction is very desirable, if at all possible, since subsequent deformation will be minimized and greater stability of the road achieved.
- The fabric is then rolled out over the site by hand (if the job is small or if equipment cannot maneuver) or using construction equipment. A front end loader can be easily fitted with a sling supporting a pipe, which is run through the core of the fabric roll. Caution must be exercised, however, if machinery is being used so that the wheels or tracks do not tear the fabric. Thus it is preferable if the working platform is the soil subgrade and not the previously laid fabric.
- The fabric must be overlapped at the end of each roll and on the sides of adjacent fabric sections. The amount of this overlap depends upon the compressibility of the soil, the nature of the traffic being supported, the thickness of the aggregate being placed on the fabric, and the deformation characteristics of the fabric itself. Thus a firm rule is difficult to state. Typical overlaps are 12 to 48 inches, depending on conditions as just mentioned. Some manufacturers state their recommended overlapping in percent of roll width for different ground conditions (e.g., see Table 4.1).
- The aggregate is then placed on the fabric. Caution should again be exercised, however, since the aggregate should not be dumped directly on the fabric, nor should the dump trucks run on the fabric.

TABLE 4.1. Fabric Overlap Recommendations[a]

CBR	Overlap (%)
20	10
15	12
10	14
8	15
6	18
4	22
2	25

[a]Source: Ref. 5.

The aggregate is to be dumped on the previously placed aggregate and spread by a bulldozer or front end loader. The blade or bucket should be kept sufficiently high so that the aggregate is not being pulled over the fabric but is being dropped over a minimum height. Naturally, the quality of the aggregate will dictate how much care is required in this part of the construction sequence.

- A road grader should rough grade the site, followed by compaction and then final grading.
- Placement of a wearing surface, if required, then follows along standard procedures.

General Behavior

CASE 1. In an extensive series of tests to investigate and test the feasibility of various new techniques for constructing bridge approach roads across soft soil, the Corps of Engineers has evaluated two types of fabrics.[7] Each test section was constructed under shelter at the Waterways Experiment Station (WES) in Vicksburg, Miss. The test road was 204 feet long and 12 feet wide with each of seven test sections approximately 30 feet long (other systems were also evaluated along with fabrics). Subgrade consisted of a heavy clay soil with an initial strength of 0.7 to 1.0 CBR for a depth of 10 inches, then the same clay with a 1.0 to 2.3 CBR for another 14 inches. Accelerated traffic tests were conducted using various loads on a 5-ton dump truck. Traffic was recorded in terms of coverages of equivalent 18,000-pound single axle, dual wheel load operations.

The control section consisted of 14 inches of crushed stone placed directly on the clay subgrade. It developed an 11-inch rut after 200 coverages of traffic.

An identical section to the control section was made, with the exception that Bidim (12 ounces per square yard) was placed beneath the stone and directly on top of the clay. This section withstood 2,500 coverages before developing an 11-inch rut. This represented a savings in design thickness of approximately 27 percent. The fabric did not tear and failure was caused by upheaval of the subgrade due to shear failure in the underlying clay soil.

Another section was constructed identical to the control section, but an impervious neoprene-coated woven nylon membrane (T-16 at 18.5 ounces per square yard) was placed between the stone aggregate and the clay subgrade. This section withstood 37,000 coverages before developing an 11-inch rut. This represented a design thickness savings of approximately 48 percent. The fabric did not tear and failure was caused by subgrade rutting but was not accompanied by base flow under the wheel paths. "Performance of this item was outstanding."[7]

CASE 2. ROAD CONSTRUCTION ON MUSKEG. A test section of road in the Tongass National Forest near Petersburg, Alaska consisted of about 10 feet of muskeg at its surface and was to be used to support logging traffic.[8] The average vane shear strength of the peat was 250 pounds per square foot and the saturated soil had a water content of approximately 960 percent. The test section was subdivided in stations along its length into zones of a double layer of fabric, a single layer of fabric, and no fabric. The fabric used was Fibretex. The test was planned to illustrate the differences in thickness of rock fill required to reach a stable and permanent road surface. The rock fill was highly variable with size ranging, from 4 feet to sand size.

Settlement and thickness measurements indicated that where no fabric had been used, depths of fill from 5 to 7.5 feet were required. With fabric (either single or double thickness) the required depth of fill was 3.5 and 5.5 feet. Thus the savings in rock fill amounted to about 28 percent. Clearly, the presence of fabric prevented local bearing capacity failures from occurring and proved the value of fabric utilization.

CASE 3. EMBANKMENTS OVER SOFT SOILS. The Netherlands State Road Laboratory has examined the function of fabrics below sand embankments in seven 120 by 50 meter areas.[9] The subgrade soil was 0.40 meter of clay underlain by 4.0 meter of peat, below which was a firm sand. The potential failure surface was a circular arc, typical of many deep-seated base failures of embankments. The main purpose of the testing was to see what height of sand fill was required to bring about instability and to assess the interaction of the fabric. Where failure did not occur the tension in the fabric was measured and compared to design criteria.

Discussion in the paper of the stress-strain characteristics of the fabric vis-a-vis the load/determination behavior of the soil mass and of the effects of fabric prestress is significant. A most important finding of the study is that a reinforcing fabric membrane will only improve stability if it is able to develop much higher strength by substantially smaller deformations than the soils acting alone.

It is critical in this type of application that the strength actually be mobilized. Therefore, the fabric may require some type of anchoring. Broms[10] has proposed a number of possible schemes in this regard (see Figure 4.7).

CASE 4. CONSTRUCTION ACCESS ROADS. In order to stabilize about 5 miles of access and haul roads and 180 acres of parking, over one million square yards of fabric are being used between the soil subgrade and 12 to 18 inches of stone aggregate surfacing.[11] Almost all of the fabric is Mirafi 140. The site is near Rockport, Indiana along the Ohio River and is for an electric generating plant of the Indiana and Michigan Electric Co.

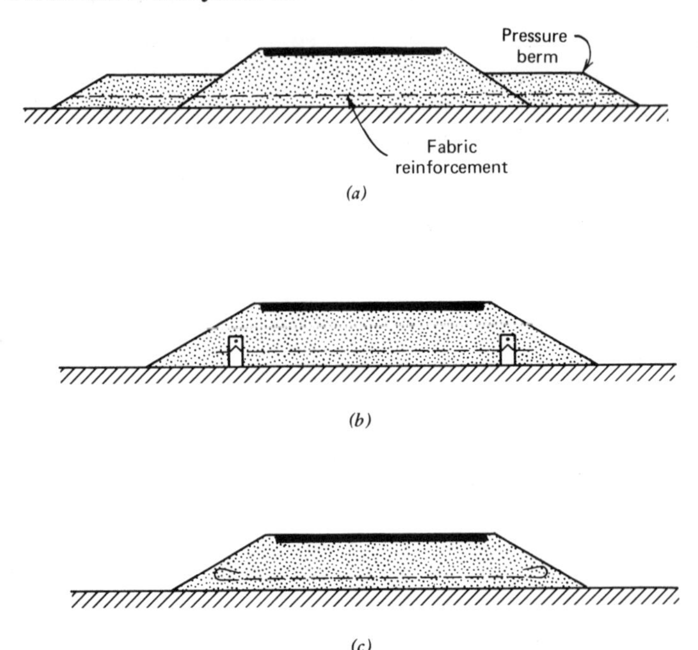

FIGURE 4.7. Anchor systems proposed by Broms (from Ref. 10). (*a*) Pressure berms; (*b*) anchor blocks; (*c*) folding of fabric.

Installation is as follows: after an initial clearing to remove large foliage, the fabric is unrolled by laborers and laid flat on the subgrade. More than 300 linear feet are on each of the rolls, which weigh about 165 pounds and are 14.7 feet wide. Where 2 or more sheets of fabric are needed along side of each other a 9-inch overlap is used.

From 12 to 18 inches of minus 1-inch initial run crushings are end dumped on the fabric, making sure that the haul truck is operating on at least 12 inches of previously placed aggregate. It is then spread by a small dozer and compacted with 2 to 6 passes of a Raygo 13-ton vibrating roller. The roadway surface is then immediately opened to haul traffic. Results to date have been very satisfactory.

CASE 5. BEARING CAPACITY OF SHALLOW FOUNDATIONS. Although we know of no specific case histories where fabrics have been used to support shallow foundations as such, there is no reason why fabrics cannot be used in marginal situations to help support footing loads. In cases where the subsoils have inadequate shear strength characteristics, both cohesion and friction can be improved by proper placement of the fabric. By improving these properties of the foundation soil, the bearing capacity will be improved. This is easily seen by considering the bearing capacity formula as conventionally taught in geotechnical engineering courses,

$$p_0 = cN_c + qN_q + \tfrac{1}{2}\gamma BN_\gamma \qquad (4.3)$$

where p_0 = ultimate bearing capacity of the footing
c = cohesion
$q = \gamma D_f$ = surcharge load above footing elevation
γ = unit weight of soil
D_f = depth of footing beneath ground surface
B = width of footing
N_c, N_q, N_γ = bearing capacity factors.

The bearing capacity factors are all trigonometric functions of the friction angle ϕ of the soil. As clearly seen in Figure 4.1, fabric increases the shear strength which is, in effect, increasing c and/or ϕ, thereby increasing p_0 in equation 4.3.

Proper placement is important in this type of application, and Mitchell[12] gives some insight into potential failure mechanisms (see Figure 4.8).

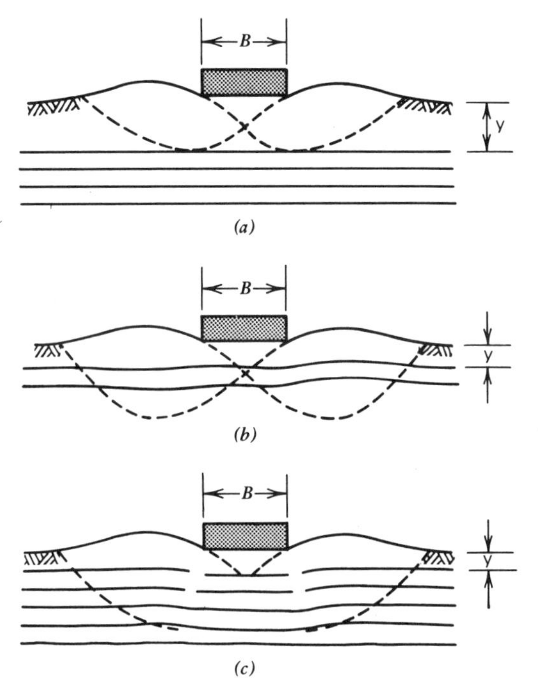

FIGURE 4.8. Possible failure modes in footings for improperly placed or inadequate fabric reinforcement (after Mitchell, Ref. 12). (*a*) $y/B > \frac{2}{3}$: shear above reinforcement; (*b*) $y/B < \frac{2}{3}$ and $N < 2$ or 3, or short ties: ties pull out; (*c*) $y/B < \frac{2}{3}$, long ties and $N > 4$: upper ties break.

4.2 Fabrics in Slope Stability Problems

Background

The failures of soil slopes and their foundations have been actively studied since the early 1900s. Many concepts and theories have evolved during the ensuing years, but most seem to revolve around the concept of a factor of safety against failure. The concept is illustrated in Figure 4.9 for a circular arc type of failure. The factor of safety is as follows:

$$FS = \sum \frac{\text{resisting moments}}{\text{driving moments}}$$

$$= \frac{\tau \widehat{ab} R}{Wx}$$

where τ = shear strength of the soil along arc length ab
R = radius distance from the failure arc to the hypothetical center of the slide
W = weight of the soil mass involved in the potential slide
x = horizontal distance from the center of gravity of the potential sliding mass to the hypothetic center of the slide.

Upon finding the minimum numeric value of the factor of safety (this requires a search of all possible radii and coordinate locations of the hypothetical center of the slide), it is then compared to unity. If $FS > 1.0$ the slope is stable (its relative stability depending on the actual numeric

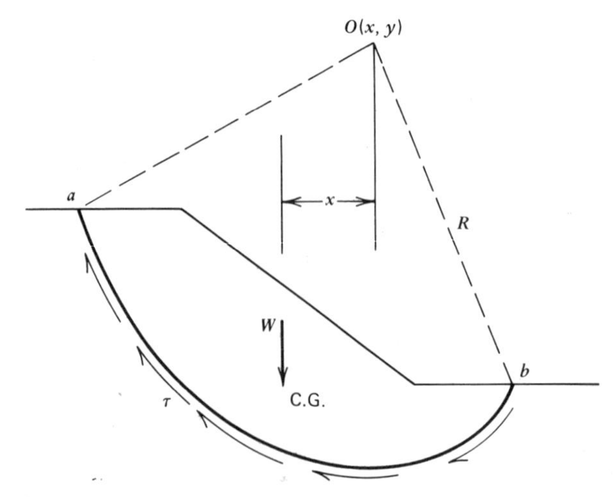

FIGURE 4.9. Schematic diagram of slope stability failure.

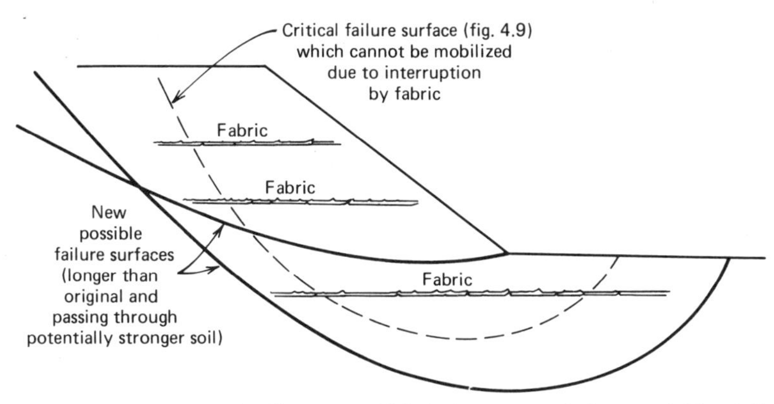

FIGURE 4.10. Schematic diagram of fabric-interrupted slope stability failure.

value, e.g., a slope with $FS = 2.0$ is quite safe whereas one with $FS = 1.1$ is questionable), if $FS \simeq 1.0$ the slope is on the verge of failure (often called incipient failure), and if $FS < 1.0$ failure will result.

The role that fabrics can play in affecting slope stability can easily be seen in Figure 4.10. By placement of fabric across the potential failure plane the arc length *ab* is changed and increased, thereby increasing the factor of safety. The new failure plane must move around the fabric into less critical failure surfaces than the most critical failure path without the fabric reinforcement. The tensile strength of the fabric is mobilized in this type of situation, during which the fabric will elongate and deform. Thus sustained load or creep strength appears to be of importance in transferring the load from the soil to the fabric, then back to the soil again, but at a different location.

This type of soil/fabric interaction holds great promise for the use of construction fabrics in problems involving slope stability.

Case Histories

CASE 1. FABRIC REINFORCEMENT AGAINST SLOPE INSTABILITY. Fabric has been used near Gothenburg, Sweden to increase the stability of an embankment behind a bridge abutment that was supported on point-bearing timber piles.[10] As shown in Figure 4.11, the embankment is a rock fill, beneath which is three layers of polyester fabric. The layers were placed 30 centimeters apart. The site was completely instrumented via inclinometers, piezometers, and earth pressure cells and was also monitored by surveying. The lateral displacements have been relatively small.

CASE 2. CONTAINMENT DIKES FOR DREDGED MATERIAL. Although it has just recently been completed and no field data is yet available, the advanced

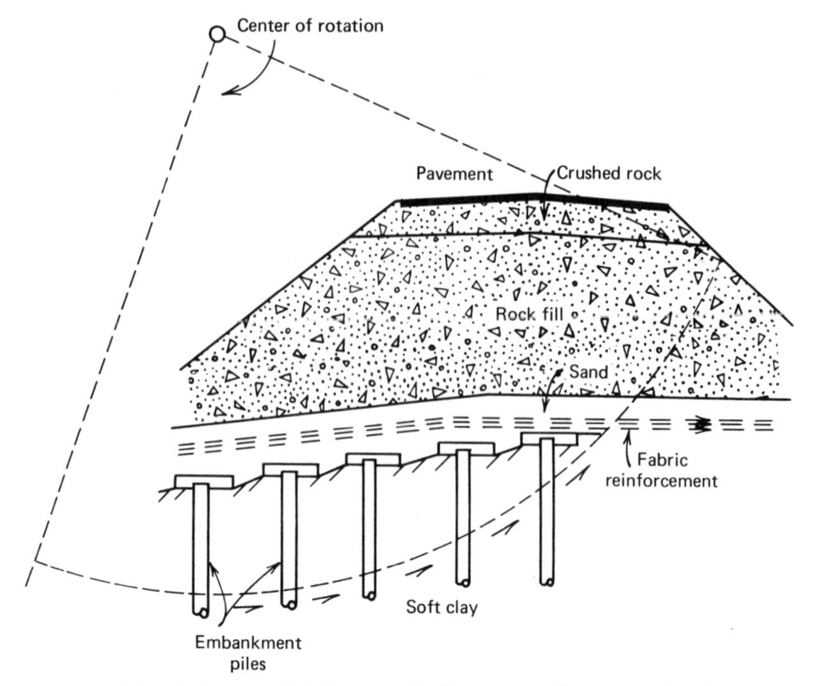

FIGURE 4.11. Polyester fabric as reinforcement in an enbankment near Gothenburg, Sweden (after Broms, Ref. 10).

planning and design work that has been undertaken on the use of fabric-reinforced containment dikes is worthy of note. Haliburton et al.[13] have conducted such a study for the U.S. Army Engineer District in Mobile, Alabama to obtain a long-term confined disposal capacity for dredged material taken from the Mobile Harbor. The proposed site is southeast of the City of Mobile across the Mobile River on Pinto Island. Average sediment dredging volume in the area is estimated to be about 2.0 million cubic yards within a 24-month period. This high volume of hydraulically dredged fine-grained soil is to be deposited on Pinto Island, which is underlain by high-water-content fine-grained soils at an elevation at, or slightly above, sea level.

The proposed solution by Haliburton et al. is to form containment dikes using fabrics that will separate individual zones of the embankment so as to:

- Contain the fill.
- Properly load the subsoil.
- Reinforce the subsoil.
- Provide anchorage for the fabric.

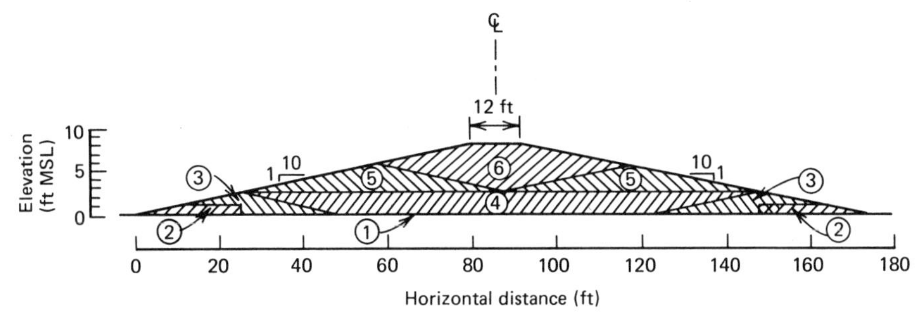

FIGURE 4.12. Cross section of fabric-containment dike near Mobile, Alabama, to contain dredge spoil. Numbers show sequence of operations (after Haliburton, et al., Ref. 13).

Figure 4.12 shows a proposed cross section of the dike. Upon completion of the dikes the area enclosed within them will be used for storage of additional dredged materials.

The required construction sequence is shown in Figure 4.12 and is extremely important in obtaining satisfactory performance of the dike. It is summarized as follows:

1. Filter cloth is laid on the surface in continuous transverse strips over a thin-sand working table, with approximately a 20-foot overlap or excess at each end. Adjacent transverse strips are slightly overlapped and sewn together.
2. As the filter cloth is placed, two outside access and anchorage strips are constructed up to about elevation 1.5, by placing material onto the filter cloth. These access strips are carried the entire length of the proposed section with the excess filter cloth at each end lapped and buried before the next operation is started.
3. Two small outside dikes are then constructed to anchor the filter cloth, and the resulting vertical settlement under these dikes stretches the filter cloth in the center section.
4. The center section is then filled to anchor the filter cloth along the entire transverse length of the dike section.
5. Intermediate dike sections are then constructed to cause settlement toward the outside of the dike, again tensioning the filter cloth in the center.
6. The center section is constructed last, to design elevation 8.

Critical in the performance of such a system is the choice of fabric wherein high tensile strength and minimum creep deformation are necessary fabric properties. Since there is a scarcity of information available in the open literature on these properties in a direct comparative situation, that is, one fabric versus another, Haliburton et al.[14] undertook such a task. Twenty-eight commercially available fabrics were tested and evaluated for suitability as dike reinforcement material.

The initial tests were strip tensile tests and those fabrics meeting the design requirements were then tested in a sustained load mode, that is, they were creep tested, tested for soil-fabric frictional resistance, and for the effects of immersion and water absorption on tensile strength. After an extended series of tests, five fabrics were recommended for consideration as dike reinforcement for this particular application.

4.3 Fabrics in Retaining Walls

Background

Other than height, the design of a retaining wall depends more heavily on the type and nature of the backfill soil than on any other factor. Design is somewhat standardized and is treated in a straightforward manner in Terzaghi and Peck.[15] Using Rankine earth pressure theory (although technically not the most accurate theory but widely used in practice) the earth pressure is considered to be linearly increasing with depth, as shown in Figure 4.13. The values for the active earth pressure as a function of depth and the resulting force for a wall of height H are

For granular soils (gravels and sands)

$$p_a = \gamma z K_a \tag{4.4}$$

$$P_a = \frac{1}{2}\gamma H^2 \tan^2 (45 - \phi/2) \tag{4.5}$$

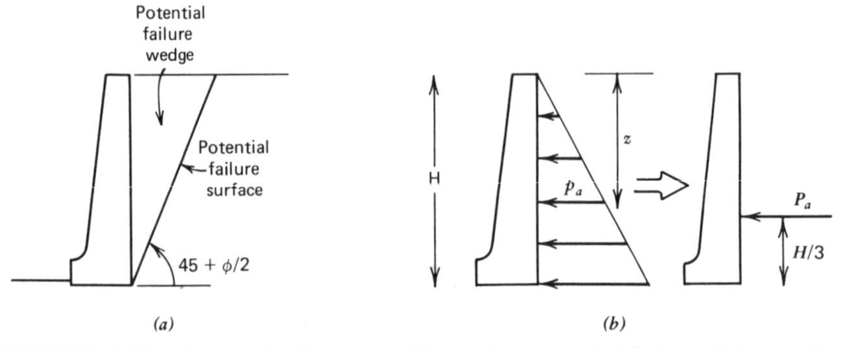

FIGURE 4.13. Schematic diagrams of lateral pressure behind retaining walls, showing (*a*) theoretical failure plane and (*b*) stress distribution and resultant earth pressure force.

For cohesive soils (silts and clays)

$$p_a = \gamma z K_a - 2c\sqrt{K_a} \tag{4.6}$$

$$P_a = \tfrac{1}{2}\gamma H^2 \tan^2 (45 - \phi/2) - 2cH \tan (45 - \phi/2) \tag{4.7}$$

The insertion of fabrics into the soil mass producing these pressures and forces is of great interest. This interaction takes the form of two separate mechanisms.

1. By intercepting the potential failure surface shown in Figure 4.10(a), the fabric causes the surface to be diverted, moving it closer to the wall, and thus less soil is involved, resulting in lower pressures.
2. By reinforcing the backfill soil, its shear strength (consisting of the cohesion c and the friction angle ϕ) is significantly increased. As seen in the equations, larger c and ϕ values produce lower earth pressures. This behavior has been shown in the laboratory by Broms,[2] who tested fabric-reinforced sands in the loose and dense state at different confining pressures and found a doubling of strength, as shown in Figure 4.1.

Case Histories

CASE 1. COMPLETE FABRIC RETAINING WALLS. A fabric retaining wall consists of horizontal layers of fabric-containing earth fill with one side of the fabric folded back around the exposed edge and lapped over the next layer to form the wall face (see Figure 4.14). At the face of the wall the fabric retains the soil from spilling out, while in the interior of the fill fabric tensile stress is developed, which stabilizes the mass. This tensile stress is developed through friction with the soil particles above and below the fabric. Four cases have been reported: one in France[16] in 1971 and three in U.S. National Forests in 1974 and 1975.[17-19] Two of these walls were instrumented to measure fabric stress, movements within the soil mass, and environmental effects of the fabric.

The Olympic National Forest Wall[18] is 166 feet long, 18.5 feet high at its maximum, and is constructed using Bidim (C-28 and C-38) and Fibretex (12.4 and 17.7 ounces per square yard). Construction was similar to a standard reinforced earth wall[20] with the exception that a temporary form was used to hold the end of the fabric (the 3-foot overlap) during placement of the soil filling each particular layer. After each soil layer was placed, the fabric overlap was folded over the fill, the layer was compacted, the form removed, and then moved upward for the next layer. Upon completion, the exposed wall face was coated with an asphalt emulsion to avoid ultraviolet light deterioration.

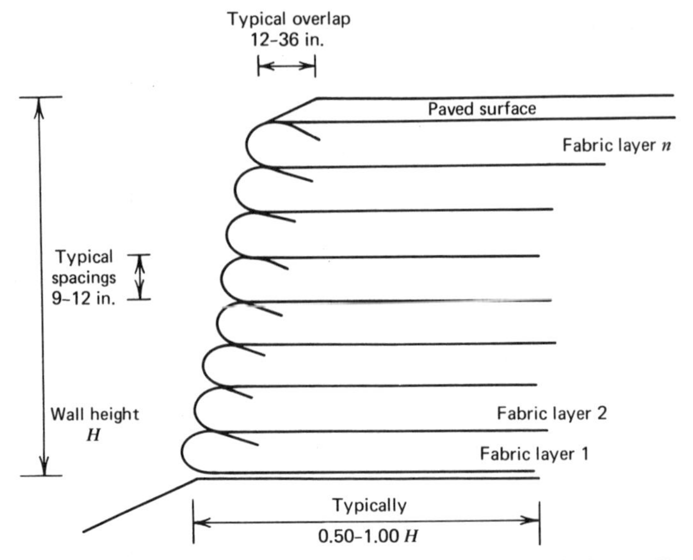

FIGURE 4.14. Typical cross section of fabric retaining wall.

Instrumentation for this particular wall included vertical and horizontal inclinometers, settlement meters, and standard surveying techniques. The results indicate vertical movements of 0.1 foot or less, with the exception of the very top of the wall. These upper movements were larger but may have inadvertently occurred during construction. Horizontal movements varied from a maximum of 0.17 foot at the wall face to zero within the wall section.

The exposed surface of fabric retaining walls is a potential problem due to uv deterioration or vandalism, and for these reasons the fabric wall at the Siskiyou National Forest was protected with gunite.[19] This wall was approximately 10 meters long and 3.3 meters high. The fabric used was Fibretex, and the wall was apparently satisfactory for the traffic loads imposed.

Another application, near Shelton, Washington, used two different fabrics (Fibretex and Bidim) and was considerably larger.[19] The wall was 60 meters long and 6 meters high at its maximum. The exposed portion of this wall was sprayed with an asphalt emulsion in the amount of one liter per square meter. It was felt that this gave adequate protection against deterioration and yet allowed for sufficient permeability of the fabric, an important consideration. This wall was instrumented with inclinometers and slip rings for measuring vertical and horizontal deformations. During the $1\frac{1}{2}$ years after its construction it has performed satisfactorily. Of interest, the authors point out that creep in the fabric is not as significant as originally considered.

A variation of this type of "complete" fabric wall has been pres
by Broms.[10] As shown in Figure 4.15, the exposed fabric at the fro
the wall is replaced by *L*-shaped concrete panels. On an equivalent
we feel that these front wall panels can be made from any one of a number of materials, for example, metal, concrete, treated wood, plastic, and so on.

CASE 2. FABRIC-REINFORCED EARTH WALLS. Similar to the previous case histories involving continuous sheets of fabric to construct earth walls and earth dams, the use of narrow fabric strips provide reinforcement in the same manner.[21] This scheme then duplicates the use of metal strips in conventional reinforced earth walls. In addition to lower cost, the fabric strips have the additional advantages of being noncorrosive under almost any environmental situation and of developing full friction between the strips and the soil.

Because of the potential influence of corrosion in metal strips, the U.S. Army Engineers Waterways Experiment Station performed a full scale test with fabric strips. Neoprene coated nylon fabric strips, 0.08 inch (0.20 centimeters) thick, 4 inches (10.2 centimeters) wide, and 10 feet (3.1 meters) long were attached to face panels as shown in Figure 4.15. The face panels (or skin elements) were of aluminum, each panel being 2 feet (0.61 meter) high, 12 feet (3.6 meters) long, and 1.6 inches (4.1 centimeters) thick. The fabric ties were fastened to the panel by bolting them to a double angle connector. The entire wall was to be 16 feet (4.9 meters) long and 10 feet (3.1 meters) high, according to the design that was

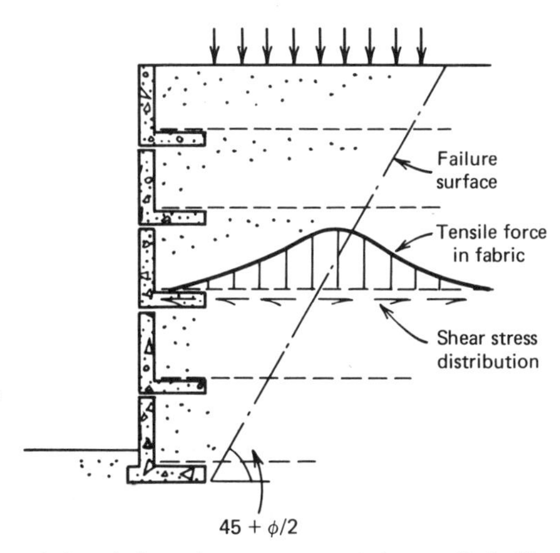

FIGURE 4.15. Fabric-reinforced concrete retaining wall, indicating shear stress distribution (from Ref. 10).

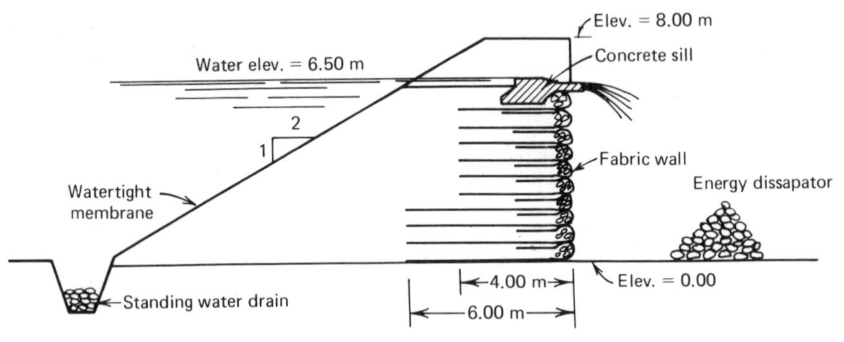

FIGURE 4.16. Schematic diagram of typical profile of fabric earth dam (from Kern, Ref. 22).

adapted directly from standard reinforced wall procedures.[20] The wall was instrumented and monitored during construction.

Failure occurred when the height of the wall was approximately 9 feet (2.7 meters). The movement of the skin element prior to failure was 2.3 inches (5.7 centimeters) at the bottom and 6.0 inches (15 centimeters) at the top. There were no visible signs of defects of the fabric reinforcement strips or the connectors. The conclusion reached by Al-Hussaini[21] is that failure of the wall was possibly caused by excessive deformations of the fabric reinforcement ties. Thus creep under sustained load appears to be the significant mechanical property of the fabric for this type of application.

CASE 3. FABRIC EARTH DAMS. While this application is similar in concept and construction method to fabric retaining walls, the hydrostatic pressure of the retained water and the potential scour problem make the application somewhat unique. Kern[22] describes an application of this type in France where a fabric wall with a vertical downstream face was constructed and used as a weir. The structure is 6.5 meters high and is capped by a concrete sill. As seen in Figure 4.16, the upstream slope is 2:1. The fabric used was a woven polyester filled with a loam soil.

4.4 Fabric Containment Systems

Background

As seen in the three previous sections, fabrics can reinforce soils by being sandwiched in layers and by supporting the soil at an exposed face. A logical extension to horizontal and vertical reinforcement is the complete encapsulation of the soil by fabric. This is the topic of this particular section.

As with most cases of reinforcement, the tensile strength of the fabric is of paramount importance. As this strength is mobilized, elongation and creep deformation under sustained loading becomes of great interest in the ultimate performance of the system.

Case Histories

CASE 1. CONTAINMENT OF LIGHTWEIGHT FILL. A series of 4- to 7-foot high dikes constructed on low-strength highly compressible marsh soils have presented continual maintenance problems since their construction and have resulted in two failures.[23] The site, containing sludge and supernatant lagoons, is located in Madison, Wisconsin. The dike rehabilitation program utilized fabrics in two different applications.

The main application was in using wooden chips as a lightweight fill and containing these by means of fabric as a form. Thus soil intrusion into the wood chips is prevented, as is lateral spreading of the wooden chips. This maintains the general configuration of the dike (see Figure 4.17). Bidim was the fabric used in this application.

A secondary application used the fabric as both a separating and a reinforcement membrane. In order to prevent the channel fill from mixing with the soft mud occupying most of the original channel, a fabric membrane was spread over the channel bottom before fill was placed. The fabric, presumably Bidim, was in 17-foot-wide (5.2 meters) panels sewn together with polyester thread with a field sewing machine. The fabric stretched significantly during channel fill placement, but did not break or tear either within the fabric itself, or at the seams.

The first phase of this project, consisting of 600 feet of dike rehabilitation, has been successfully completed. A second phase, consisting of 3,000 feet, will be undertaken shortly.

CASE 2. CONTAINMENT OF FINE-GRAINED SOIL FOR ROAD BASE. As reported by Smith and Pazsint,[24] the Corps of Engineers' Cold Regions Research and Engineering Laboratory has used fabrics to envelope fine-grained soils for the purpose of road construction. The main objectives were to utilize fine-grained soils where no granular soils are available and to build all-weather road surfaces without any overlying structural layer. This situation of a lack of granular soils is widespread in the arctic and subarctic along with the problem of permafrost in many of these same areas.

The test section was an unimproved road near Fairbanks, Alaska, which had experienced significant thaw weakening. The membrane used

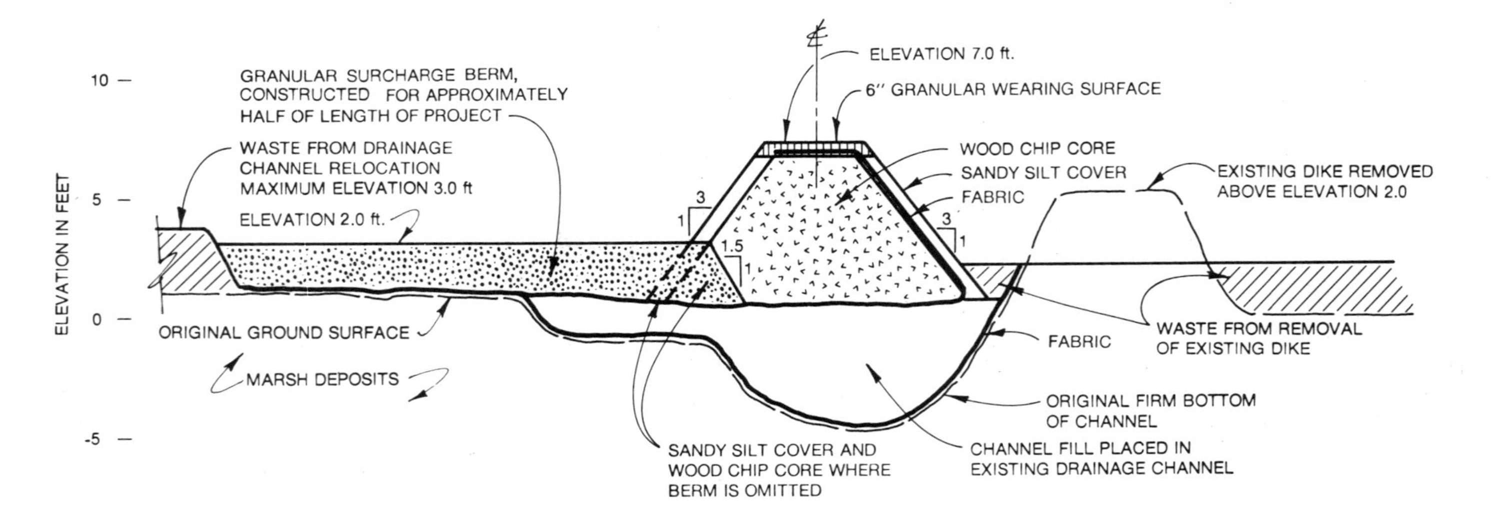

FIGURE 4.17. Typical section of demonstration project (after Roth and Schneider, Ref. 23).

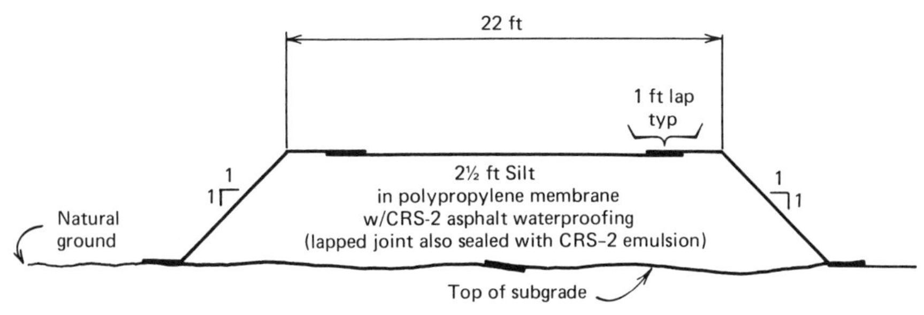

FIGURE 4.18. Schematic diagram of fabric-encapsulated soil for road bases (from Ref. 24).

to construct the section was Petromat, and the general configuration is shown in Figure 4.18. The section is approximately 28 feet wide and 260 feet long.

The subgrade was prepared and rolled and then covered with the fabric. CRS-2 emulsified asphalt was used to seal the longitudinal joint at the centerline, which was overlapped 1 foot. The entire fabric was also sealed before placing the silt, which was done with a road grader and roller in three lifts. The upper surface was then covered with fabric, as were the sides, thereby completely enveloping the silt. Sealing was done with the same emulsified asphalt. A $1\frac{1}{2}$-inch layer of gravelly sand was spread over the top surface.

After two years of service the road section has performed satisfactorily, although two problems did occur. One was an area of significant rutting, the other a section where the joint separated. Both occurred when the fabric was exposed to infiltration of surface water. Benkleman beam tests were performed to measure deflections, which were generally small. These tests suggested that no deterioration occurred with freezing and thawing periods. In general, the condition of the traffic surface was considered satisfactory for an expedient road such as this one.

A number of cautions were also pointed out. They are as follows:

- Moisture control during the placement of the enveloped silt is critical.
- The membrane must be completely impermeable to prevent water from entering the enveloped silt. If this occurred, the silt would immediately lose its strength, causing a local bearing capacity type of failure.
- Snow removal equipment can easily tear the membrane.
- Windy weather can be troublesome when large areas of the membrane are exposed.

In a parallel series of tests, Eaton[25] has reported that moisture migration in the field was almost identical to that found in the laboratory. The amount of frost heave within the encapsulated soil itself was negligible, due primarily to the low water content of the silt fill. All heave took place in the subsoils beneath the encapsulated soil, which had a higher water content and had access to water from below.

The pavement deflections as measured with the Benkleman beam test showed very small movement, which suggests that the encapsulated soil maintains its load-bearing strength during periods of thaw. Eaton concludes by suggesting that areas requiring further research are the membrane itself, sealers, chemical stabilization, and mechanical drying of the soil to be incorporated within the fabric envelope.

4.5 Control of Reflective Cracking

Background

Cracks in existing pavements often reappear in asphalt overlays unless remedial measures are taken before resurfacing. This phenomenon, called reflective cracking, is often very difficult to prevent because the original source of the cracks is usually beneath the pavement in either the stone base or in the soil subgrade. To excavate and replace the objectionable materials with suitable ones is obviously a major expenditure of time and money. Such a procedure would, in fact, completely destroy the efficiency and speed of a typical bituminous overlay project. The more expedient method of filling in the cracks with hot mix, cold patch, or liquid asphalt usually only sidesteps the problem on a temporary basis.

It is felt that construction fabrics laid on the pavement before resurfacing hold the key toward an economical method to eliminate or reduce reflective cracking from occurring. The fabric that is sandwiched between the existing pavement and the new overlay is acting in the capacity of both a separator and as a reinforcement medium. Thus this use could be included in Chapter 3 on separation using fabrics equally as well as in this section. A secondary feature is that the fabric may act somewhat in the capacity as an inhibitor of water from entering into the pavement subgrade. As shown in Chapter 5, it is vital to pavement life to keep the subgrade either free of water or adequately drained.

The major fabric properties relied upon in this particular application are strength, temperature resistance, asphalt retention, and comformability.

Construction Procedure

The construction procedure for placing fabric beneath a bituminous overlay to eliminate or control subsequent reflective cracking depends on a number of factors. The major considerations appear to be climate, nature of traffic, severity of the problem in the original pavement (and/or its subgrade), and the physical and mechanical properties of the fabric. thus each situation is somewhat unique, but there are enough similarities that a generalized procedure can be outlined.

- The existing road surface must be cleaned of dirt and vegetation and and must be dry.
- Existing cracks between $\frac{1}{8}$ and $\frac{1}{4}$ inch thick are to be filled with crack filler. Larger cracks or holes are to be filled with asphalt, hot mix, or cold patch.
- An asphaltic sealant is then uniformly sprayed over the existing pavement. The amount ranges from 0.15 to 0.40 gallons per square yard, depending on the porosity of the existing pavement and the absorbancy of the fabric. Recommended sealants are asphaltic cement (AC-10 or AC-20),[26] cationic asphalt emulsions (CRS-2 or CRS-1h)[27] and anionic asphalt emulsions (RS-2 or RS-1).[27] Caution is required when using the asphalt emulsions to ensure that they adequately cure before the fabric is placed. This takes from 30 minutes to 4 hours, depending on the temperature and humidity.[27] Cutback asphalts cannot be used with polypropylene fabrics, since the solvent in them reacts with the polymer at high temperatures.
- The fabric is placed on the sealant by hand or with mechanical equipment. Excessive wrinkles or folds in the fabric must be cut open and laid flat. Stiff brooms are used to get a good bond and smooth out the surface as well. Joint overlaps of 1 to 3 inches are generally used. Additional sealant should be applied at the joints. If sealant comes through the fabric, sand can be spread over it to absorb it.
- The hot mix overlay is placed directly on the fabric as soon as possible. It is generally not necessary to place a tack coat on the fabric. The temperature of the mix should be about 300°F with a maximum of 325°F. Care must be taken to avoid movement or damage to the fabric from turning of the paver or truck movement.
- It should be emphasized that this use of fabrics to retard or eliminate reflective cracking is for asphalt overlays on top of existing asphalt

pavements. Cracked Portland cement concrete pavements require special treatment to eliminate the causes of the cracks before applying fabric and resurfacing. A number of case histories on the use of fabrics to control reflective cracking follow.

Case Histories

CASE 1. Beginning in October 1968, thirty-seven projects aimed at evaluating fabrics in the control of reflective cracking in bituminous overlays have been initiated by the Federal Highway Administration, Department of Transportation, under its National Experimental and Evaluation Program (NEEP). The most comprehensive of these programs were conducted by Arizona, California, Florida, and North Dakota. The general finding was that there is no strong evidence that the fabric used provides a universal mechanism for extending the crack-free life of an overlay. Most of the projects that have shown reduction in reflection cracking using fabrics have been in areas with relatively mild climates.

The conclusion of the Federal Highway Administration is that fabrics may be used in the control of reflection cracking, but best results can be achieved where the existing pavement has experienced fatigue-associated alligator cracking with cracks $\frac{1}{8}$ inch wide or less and located in relatively mild climates. They also express some concern for future recycling of pavements where fabric is a part of the existing pavement.

Current Federal Highway Administration recommendations are to continue with the experimental program and to evaluate effectiveness of a wide range of fabrics on a state-by-state basis.

To be sure, the above summary statements challenge the universal use of fabrics in the control of reflection cracking in bituminous overlays. In general, however, we feel that reports are definitely on the positive side. A sampling of the many case histories that are available follows.

CASE 2. A portion of I-95 in Pittsfield, Maine was in need of repaving, and past experience had shown that a significant percentage of the cracks in the old pavement would in time reflect through the overlay.[28] In order to reduce the amount of reflective cracking, Mirafi, Typar, and Bidim C-22 and C-28 were used on different sections before placing the overlay.

The pavement was prepared by cleaning and then sprayed with an emulsion (RS-2) in the amount of 0.10± gallons per square yard. In addition to sections that were barely cracked, where fabric was used to cover the entire section of pavement, several smaller pieces of fabric were used to cover individual cracks. These smaller strips were placed over a tack coat of AC-90, and in some cases the larger cracks were filled with cold patch prior to placing the fabric.

The fabric was placed by hand with lateral spreading being done by laborers using push brooms. Any significant remaining wrinkles were split with a knife and flattened into the tack coat. The fabric was exposed for as long as a week. Just before the paver reached the fabric a few shovels of mix were placed on it to prevent the equipment from catching on the leading edge of the fabric.

No results are available as to the performance of the fabric or of comparison of one fabric to the next.

CASE 3. Of interest is an 18-mile (20-kilometer) stretch of I-20 near Peco, Texas, which is being used as a test site for the Texas Department of Highways to evaluate fabrics in resisting reflecting cracking.[29] The highway carries a daily average traffic volume of 3,450 vehicles, of which one-fourth are trucks.

Four test sections were constructed as described in Table 4.2. The installation procedure was to apply a tack coat, then the fabric (in this

TABLE 4.2. Pavement Test Section Descriptions

Section	Length[a] (miles)	Pavement Composition[b]	Pavement Thickness (in.)
1	7.5	Asphalt-crushed rock ($\frac{3}{8}$ in. max) binder; $\frac{3}{4}$-in.-thick asphalt concrete	$1\frac{1}{4}$
2	7.5	Petromat fabric; asphalt-crushed rock ($\frac{3}{8}$ in. max.) binder; $\frac{3}{4}$-in.-thick asphalt concrete	$1\frac{1}{4}$
3	1.5	2-in.-thick asphalt concrete; petromat fabric; asphalt-crushed rock ($\frac{3}{8}$ in. max.) binder; 1-in.-thick asphalt concrete	3
4	1.5	Asphalt-crushed rock ($\frac{3}{8}$ in. max.) binder; 2-in.-thick asphalt concrete; 1-in.-thick asphalt concrete	3

[a] Total length of test area is 18 miles, two lanes wide.
[b] All sections initially double bituminous surface treatment; materials applied in order listed.

case Petromat), then seal the fabric with the asphalt wearing surface. The production was 2 to 3 miles (3 to 5 kilometers) of double lane resurfacing each day. The only major problem experienced while laying the fabric resulted from western Texas' persistant winds. They tended to lift the fabric off the emulsified asphalt. Use of asphalt reduced the problem. Preliminary data[30] on the percentage of cracks reflected in the various test sections are given in Table 4.3. Here the effectiveness of the fabric versus standard and control sections can be seen.

CASE 4. The rebuilding of five miles of city streets in Kinloch, Missouri, a suburb of St. Louis, recently made use of Bidim fabric. The original roads, consisting of 2 to 3 inches of asphalt, were badly broken and showed alligator cracks and indications of subgrade water from the high-water-content clay subgrade soils. Pipes and other utilitiy lines prevented subgrade excavation and placement of granular fill materials, so fabrics were used instead.

TABLE 4.3. Performance of Test Overlays in Controlling Reflective Cracking

Test Section (Date Placed)	Cracks Reflected (%)	
	First Year	Second Year
Emulsion seal coat (8/73); $1\frac{1}{4}$ in. HMACP (4/74)	2.0	9.0
Emulsion + 3% latex seal coat (7/73); $1\frac{1}{4}$ in. HMACP (4/74)	12.0	17.0
Emulsion + 6% latex seal coat (7/73); $1\frac{1}{4}$ in. HMACP (3/74)	31.0	45.0
Emulsion + 10% latex seal coat (7/73); $1\frac{1}{4}$ in. HMACP (3/74)	14.0	19.0
Petromat, $1\frac{1}{4}$ in. HMACP (3/74)	0	1.0
$1\frac{1}{4}$ in. HMACP (3/74); Petromat, AC-3 seal coat (4/74)	0	0
Petromat, emulsion seal coat (7/73); $1\frac{1}{4}$ in. HMACP (3/74)	0.5	11.0
Petromat, emulsion seal coat (7/73)	77.0	77.0
$2\frac{1}{2}$ in. HMACP (4/74)	1.5	2.0
$1\frac{1}{4}$ in. HMACP, Petroset emulsion (4/74)	25.0	33.0
$1\frac{1}{4}$ in. HMACP (3/74) (control section)	19.0	40.0

The original design called for a 7-inch base course and asphalt paving, which was replaced with a 3-inch base course, fabric, and then asphalt paving directly on the fabric. One section was excavated and showed no signs of deterioration from the direct paving method. A 20-year paving life using fabric, versus a 5- to 10-year life using standard methods, is anticipated by the project engineers.

CASE 5. In a related problem of pavement deterioration, the Maine Department of Transportation has used fabrics to prevent the upward pumping of silt through pavement during spring thaw,[28] ("frost boils"). In a series of experimental efforts on Route 150 in Parkman, Maine it was found that the placement of fabric, in this case Bidim, on top of the soil subbase and beneath the clean gravel was successful in preventing the silt from surfacing. Other attempts were made at placing the fabric directly on the deteriorated pavement, then the gravel and a new asphalt surface on top. This approach was apparently also successful. Test holes were excavated and no silt was noticeable on top of the fabric or in the gravel, but upon removal of the fabric the underlying soil was seen to be covered with a thin layer of saturated fines. These fines had apparently been stopped from upward migration by the fabric, which acted as a separating filter material.

4.6 Fiber- and Fabric-Reinforced Concrete

Background

Historically fibers have been used to reinforce many different types of building materials. Some classic uses are straw in bricks, animal hair in plaster, and asbestos in cement. More recently, however, fiber-reinforced concrete has concentrated on steel, glass, and plastic fibers being placed in mortars and concretes to improve their material characteristics, particularly those of flexural and impact strength. An overview of these fibers is given in Table 4.4.

The steel fibers are either round (made from wire) or rectangular (made from shearing sheets or from flattened wire) with lengths of 0.25 to 3 inches. Aspect ratios (length to thickness) range from 30 to 150. Smooth, crimped, and deformed fibers have been used.

Typical glass fibers have diameters ranging from 0.2 to 0.6 mils, but are usually bonded together to form diameters of 0.5 to 50 mils. Lengths are generally from 0.5 to 2.0 inches.

Typical plastic fibers that have been used are nylon, polypropylene,

TABLE 4.4. Typical Properties of Fibers

Fiber	Tensile Strength (psi)	Young's Modulus (psi)	Ultimate Elongation (%)	Specific Gravity
Acrylic	$30–60 \times 10^3$	0.3×10^6	25–45	1.1
Asbestos	$80–140 \times 10^3$	$12–20 \times 10^6$	~0.6	3.2
Cotton	$60–100 \times 10^3$	0.7×10^6	3–10	1.5
Glass	$150–550 \times 10^3$	10×10^6	1.5–3.5	2.5
Nylon (high tenacity)	$110–120 \times 10^3$	0.6×10^6	16–20	1.1
Polyester (high tenacity)	$105–125 \times 10^3$	1.2×10^6	11–13	1.4
Polyethylene	$\sim 100 \times 10^3$	$0.02–0.06 \times 10^6$	~10	0.95
Polypropylene	$80–110 \times 10^3$	0.5×10^6	~25	0.90
Rayon (high tenacity)	$60–90 \times 10^3$	1.0×10^6	10–25	1.5
Rock wool (Scandinavian)	$70–110 \times 10^3$	$10–17 \times 10^6$	~0.6	2.7
Steel	$40–600 \times 10^3$	29×10^6	0.5–3.5	7.8

Source: Ref. 31.

polyethylene, polyester, and rayon. Fiber diameters of 0.8 to 15 mils have been used, with lengths of 0.5 to 2.0 inches being customary.

In general, the addition of fibers to concrete results in the following improvements:

- Greater resistance to cracking.
- Greater resistance to thermal changes.
- Thinner design sections.
- Less maintenance.
- Longer life.

These items can be appreciated better by noting the advantages listed in Table 4.5 for the addition of fibers to concrete.

Regarding the amount of the fibers to add to concrete, little by way of standardization is available. The criterion is often dictated by how much fiber can be added before the mix becomes unworkable. This depends not only on volume, aspect ratio, type, and kind of fiber but also on aggregate size, amount of sand, and amount of cement. Hoff[32] recommends $\frac{3}{8}$-inch maximum aggregate size, using more sand than normally required; aggregate ratios between 50:50 and 70:30, and high cement contents (up to 850 pounds per cubic yard).

TABLE 4.5. Plain Concrete Versus Fiber-Reinforced (Steel) Concrete

Property or Index	Advantage of Fibrous Concrete Over Plain Concrete (Times Higher)
First-crack flexural strength	$1\frac{1}{2}$
Ultimate modulus of rupture strength	2
Ultimate compressive strength	$1\frac{1}{4}$
Ultimate shear strength	$1\frac{3}{4}$
Flexural fatigue endurance limit	$2\frac{1}{4}$
Impact resistance	$3\frac{1}{4}$
Sand blast abrasion resistance index	2
Heat spalling resistance index	3
Freeze-thaw durability index	2

Source: Ref. 31.

Construction with Fiber-Reinforced Concrete

All types of mixers have been used successfully to mix fibers in concrete, but utmost care must be exercised to prevent balling of the fibers. If possible, the fibers should be preblended with the aggregate or added together with the aggregate. Balling is a function of a number of variables, including type of fiber, length of fiber, and aspect ratio, in that order. If balling of the fibers in the final concrete is allowed to occur, the results are generally unsatisfactory insofar as strength and other properties are concerned.

Steel fibers for reinforcement of concrete are by far the most popular type. The amount of fibers that can be added is a function of the workability, which in turn is dictated by the particular application. In general, a fiber-reinforced concrete requires more vibration to move the material and for ultimate consolidation as well. Steel fiber concrete has seen its major applications in airport and highway paving, overlays, and patching. Other applications are precast pipe, refractory concretes, architectural concrete, concrete erosion protection structures, burial vaults, and machine bases.

Glass fibers in concrete were initiated in the 1950s but were largely unsuccessful due to their loss of strength in the high alkaline environment of Portland cement concrete. This resulted in an alkali-resistant glass fiber produced in England by Fiberglass, Ltd. and in the United States by Owens-Corning Fiberglas Corp.[32] Using such fibers, typical

TABLE 4.6. Plain Concrete Versus Fiber-Reinforced (Glass) Concrete

Strength Property	Strength Ratio of Glass-Fiber-Reinforced Concrete Compared with Unreinforced Concrete (Volume Percent of Fibers)			
	0.5	1.0	1.5	2.0
First crack	1.22	1.42	1.63	1.85
Ultimate	1.25	2.00	2.75	3.50
Impact	2.3	3.6	4.9	6.2

Source: Ref. 32.

improvements as shown in Table 4.6 were measured. The mix was at a water:cement ratio of 0.45 using 0.5-inch-long fibers.

Applications of glass fibers are in precast concrete blocks, architectural products (e.g., shingles and siding panels), precast pipe, patching materials, and in airport and highway paving.

Plastic fiber introduction into concrete began in the middle 1960s with the objective of increasing spall and impact resistance. Nylon, polypropylene, polyethylene, Saran®, Rayon® acetate, Orlon®, and Dacron® have been studied. Generally the fibers varied in diameter from 3 to 630 denier and were of 3-inch length or shorter. Volume percents varied from 3 to 7 percent. All increased the viscosity of the mix but produced varied results on static strength. The fibers do have the effect, however, of holding cracked pieces together once failure begins, thereby improving impact resistance.

Case Histories

The use of plastic fiber reinforcement in building materials is still in its formative stage and little is available except for the following:

CASE 1. Hoff[32] reports that nylon fibers have been used as a filler in cement pastes in grouting of deep well casings.

CASE 2. A patented process by the Shell Chemical Company[33] uses polypropylene fibers in concrete, called Caricrete® The fibrillated film fibers are being used as concrete reinforcement. These fibers are from 6,000 to 26,000 denier, ranging from 0.5 to 4 inches length and in weight concentrations of 0.2 to 0.5 percent. Panels, slabs, piles, flotation units, and swimming pool linings have been made with these fibers. Results from impact tests show the improved properties (see Table 4.7), which are also seen in the flexural strength tests of Table 4.8.

TABLE 4.7. Impact Tests on Slabs 0.7 × 1.5 Meters, with Mortar Ratio of 1 : 3

Test	Thickness of Slab (mm)	Distance at Which First Failure Occurs (m)		Type of Failure	Remarks
		Drop Height	Horizontal Distance		
Falling steel ball (1 kg)					
Control	20	0.50	—	Completely broken	Slab broke in direction of width into two equal pieces
Caricrete	20	0.70	—	Cracked	After appearance of crack a deflection of 8 mm was measured in the center of the slab; during continuation of the test up to a drop height of 2.5 m; this deflection increased to 20 mm while a dent was formed at the point of impact; however, the slab did not break
Pendulum (25-kg sandbag)					
Control	40	0.27	1.25	Completely broken	Slab broke along a line through the center and parallel to the two
Caricrete	40	10.2	2.25	Cracked	Three cracks over the whole width. During repeated tests up to the maximum height of 3 m the slab remained in one piece. At the end of the series of tests the slab was still flat.

Source: Ref. 33.

TABLE 4.8. Bending Strength of Beams 0.1 × 0.1 × 0.3 m[a]

Mortar				Concrete			
		Bending Strength (MN/m²)				Bending Strength (MN/m²)	
Mix[b]	Density (kg/cm³)	I	F	1 : 2 : 4 Mix	Density (kg/cm³)	I	F
A Control	2.12	2.93	2.93	Control	2.36	4.71	4.71
	2.11	2.67	2.67		2.35	4.40	4.40
	2.11	2.75	2.75		2.35	3.86	3.86
Average	2.11	2.78	2.78	Average	2.35	4.32	4.32
A Caricrete	2.10	3.66	5.42	Caricrete	2.32	5.24	6.48
	2.10	3.91	7.00		2.32	4.19	5.36
	2.10	3.94	8.00	(0.5% weight)	2.33	4.59	5.26
Average	2.10	3.81	6.80	Average	2.32	4.67	5.70
A Caricrete	2.08	5.28	6.51				
	2.08	4.28	7.95				
	2.07	5.46	7.64				
Average	2.08	5.00	7.69				
B Control	2.03	1.85	1.85				
	2.03	2.35	2.35				
	2.02	2.20	2.20				
Average	2.03	2.13	2.13				
B Caricrete	1.98	2.97	4.60				
	1.99	2.38	4.81				
(0.8% weight)	1.98	2.47	4.97				
Average	1.98	2.61	4.80				

Source: Ref. 33.

[a] 1 kg/cm³ = 62.5 lb/ft³; 1 MN/m² = 10.2 kgf/cm² = 145 lb-ft/in².

[b] A—1 : 3 mix; B—1 : 4.5 mix.

CASE 3. In another case, rather than use separated discrete sized fibers for reinforcement, duPont[34] has experimented with their nonwoven fabric Typar as a reinforcement material for concrete. Placing a sheet of fabric one-third of the distance up from the bottom of a 12 by 12 by $\frac{3}{4}$ inch thick concrete tile, impact strengths were measured against control tiles. It was found that 5 foot-pounds were required to break the unreinforced section, while 20 foot-pounds were required for 2 ounce per square yard Typar-reinforced tiles and 30 foot-pounds were required for 3 ounce per square yard Typar-reinforced tiles. Furthermore, the pieces held together after cracking with the reinforced tiles. Regarding flexural strength, increases of 4 to 19 percent were noted, depending upon the flexural strength of the concrete itself.

4.7 References

1. E. Leflaive and J. Puig, "The Use of Textiles in Embankment and Drainage Works," *Bull. Liaison Lab. Ponts Chauss.* No. 69, pp. 61-79.
2. B. B. Broms, "Triaxial Tests with Fabric-Reinforced Soil," *C. R. Coll. Inst. Soils Text.*, Vol. III, 1977, pp. 129-133.
3. D. W. Taylor, *Fundamentals of Soil Mechanics,* Wiley, New York, 1948.
4. W. K. Beckham and W. H. Mills, "Cotton-Fabric-Reinforced Roads," *Eng. News—Record,* Oct. 3, 1935, pp. 453-455.
5. "A Method for Constructing Aggregate Bases Using 'Typar' Spunbonded Polypropylene," E. I. duPont de Nemours, Wilmington, Del., unpublished report.
6. "Thickness Design Manual (MS-1)," 8th ed., Aug. 1970, The Asphalt Institute, College Park, Md. 20740.
7. S. L. Webster and J. E. Watkins, "Investigation of Construction Techniques for Tactical Bridge Approach Roads Across Soft Ground," Tech. Report No. S-77-1, U.S. Army Waterways Experiment Station, Vicksburg, Miss., Feb. 1977.
8. J. R. Bell, D. R. Greenway, and W. Vischer, "Construction and Analysis of a Fabric Reinforced Low Embankment," *C. R. Coll. Inst. Sols Text.*, 1977, Vol. I, pp. 71-75.
9. A. C. Maagdenberg, "Fabrics Below Sand Embankments Over Weak Soils, Their Technical Specifications and Their Application in a Test Area," *C. R. Coll. Int. Sols Text.*, 1977, Vol. I, pp. 77-82.
10. B. B. Broms, "Polyester Fabric as Reinforcement in Soil," *C. R. Coll. Int. Sols Text.*, 1977, Vol. I, pp. 129-135.
11. ———, "Construction Fabric Keeps Earthmoving Project Moving," *Highway Heavy Constr.*, Feb. 1978, pp. 30-31.
12. J. K. Mitchell, "Ground Reinforcement Techniques—Overview," ASCE National Capital Section Seminar on Ground Reinforcement, George Washington Univ., Washington, D.C., January 26, 1979.

13. T. A. Haliburton, P. A. Douglas, and J. Fowler, "Feasibility of Pinto Island as a Long-Term Dredged Material Disposal Site," Misc. Paper D-77-3, Office, Chief of Engineers, U.S. Army, Washington, D.C., Dec. 1977.
14. T. A. Haliburton, C. C. Anglin, and J. D. Lawmaster, "Selection of Geotechnical Fabrics for Embankment Reinforcement," Contract No. DACW01-78-C-0055 for U.S. Army Engineer District, Mobile, Alabama, School of Civil Engineering, Oklahoma State Univ., Stillwater, Okla., May 1978.
15. K. Terzaghi and R. B. Peck, *Soil Mechanics in Engineering Practice*, Wiley, 2nd ed., 1967.
16. J. Puig, J. C. Blivet, and P. Pasquet, "Earth Fill Reinforced with Synthetic Fabric," *C. R. Coll. Int. Sols Text.*, 1977, Vol. I, pp. 85-90 (in French).
17. J. R. Bell, N. Stilley, and B. Vandre, "Fabric Retained Walls," in Proceedings of the 13th Annual Engineering Geology and Soil Engineering Symposium, Moscow, Idaho, April 1975, pp. 271-287.
18. J. Mohney, "Fabric Retaining Walls—Olympic N. F.," *Highway Focus,* Vol. 9, No. 1, May 1977, pp. 88-103.
19. J. R. Bell and J. E. Steward, "Construction and Observations of Fabric Retained Soil Walls," *C. R. Coll. Int. Sols Text.*, 1977, Vol. 1, pp. 123-128.
20. K. L. Lee, B. D. Adams, and J. J. Vagneron, "Reinforced Earth Retaining Walls," *J. Soil Mech. Found. Div. ASCE,* Vol. 99, No. SM10, Proc. Paper 10068, October 1973, pp. 745-764.
21. M. M. Al-Hussaini, "Field Experiment of Fabric Reinforced Earth Wall," *C. R. Coll. Int. Sols Text.*, 1977, Vol. I, pp. 119-121.
22. F. Kern, "An Earth Dam with a Vertical Downstream Face Constructed Using Fabrics," *C. R. Coll. Int. Sols Text.*, 1977, Vol. I, pp. 91-74 (in French).
23. L. H. Roth and J. R. Schneider, "Dike Rehabilitation Using Fabric Reinforcement and Lightweight Fill," *Highway Focus,* Vol. 9, No. 1, May 1977, pp. 17-42.
24. N. Smith and D. A. Pazsint, "Field Test of a MESL (Membrane-Enveloped Soil Layer) Road Section in Central Alaska," Technical Report No. 260, U.S. Army Corps of Engineers, CRREL, Hanover, N.H., July 1975.
25. R. A. Eaton, "Performance of Membrane Encapsulated Soil Layer Test Sections During Three Artificial Freeze-Thaw Cycles," Internal Report No. 469, Corps of Engineers, U.S. Army Cold Regions Research and Engineering Laboratory, Hanover, N.H., July 1975.
26. "Application of Typar for Control of Reflective Cracking," Internal Report, R. D. H., Jan. 3, 1978, E. I. duPont de Nemours and Co., Inc., Wilmington, Del.
27. "The Petromat System," Philips Fibers Corp., Greenville, S.C., 1978.
28. G. Baker, "Maine Department of Transportation Experience with Filter Fabrics," *Highway Focus,* Vol. 9, No. 1, May 1977.
29. P. H. Coleman, "Texas Uses Fabric to Protect Pavement from Cracks," *Civil Eng. ASCE,* Dec. 1977, pp. 74-75.
30. ———, "Can Fabrics Soften Paving Problems?," *The American City and County,* Nov. 1977, pp. 69-70.

31. G. B. Batson, "Introduction to Fibrous Concrete," in Proceedings of the CERL Fibrous Concrete Conference, Champaign, Ill., May 1972, pp. 1-25.
32. G. C. Hoff, "Research and Development of Fiber Reinforced Concrete in North America," Misc. Paper No. C-74-3, U.S. Army Engineer Waterways Experiment Station, Vicksburg, Miss., Feb. 1974.
33. "Chemicals in Building," CB 68/106, 2nd ed., April 1969, available from Shell Chemical Co., Plastics and Resins Div., 113 West 52nd Street, New York, N.Y. 10019.
34. S. Sands, "Concrete Reinforced with Typar Spunbonded Fabric," E. I. duPont de Nemours and Co., Inc., Wilmington, Del. 19898, Presented at American Concrete Institute Meeting, New Orleans, La., Oct. 20, 1977.

5

Fabric Use in Drainage

When Terzaghi developed the use of successively finer or coarser drainage layers (filters) to prevent one soil from moving with respect to another, while conducting flowing water, he provided

- A technically correct method to solve many drainage problems.
- A difficult physical configuration to construct under many circumstances.
- An ever increasingly expensive technique due to its being very labor intensive and requiring different types of granular soils and relatively large amounts of each different type.

Fabrics have already made significant strides into changing this procedure of using graded filters. In addition to largely eliminating the above objections, other interesting spinoffs have resulted. For example, less excavation, less chance for trench cave-ins, faster construction times, and so on, have resulted.

Four areas are discussed in this section: highway underdrains, drainage in earth dams, drainage behind retaining walls, and the use of fabric drains to accelerate settlement.

5.1 Pavement Underdrain Systems

Background

The useful life of a highway or airfield pavement is critically tied to the ability of the underdrain system to adequately transport and remove free water that remains beneath the pavement. Cedegren[1] shows in Figure 5.1

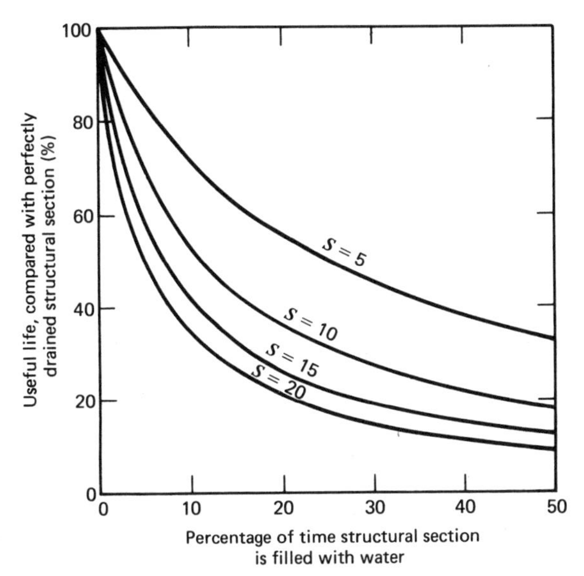

FIGURE 5.1. Effect of free water on pavement life. S is the severity factor (after Cedegren, Ref. 1).

the rapid loss of pavement life as a function of the time that water remains under the structural section. The decrease in life is greatest for high values of severity factor (S), which is the relative rate of damage per load application. The removal of free water is usually accomplished by means of a pavement underdrain system, which is fed directly from the stone base course beneath the pavement or directly from the pavement surface.

The conventional pavement underdrain system consists of a perforated pipe collector, a gravel layer around the pipe to prevent its clogging, and a sand filter layer around the gravel to prevent migration of the natural soil from entering the gravel. This system is shown in Figure 5.2(*a*).

A major use of fabrics has been to vastly alter this type of underdrain system. Shown in Figure 5.2(*b*) are the schematic diagrams of a replacement to the conventional method using a fabric-wrapped stone aggregate (with or without the perforated pipe) placed directly in an excavated trench. It has the following major advantages.

- Elimination of the filter sand in the dual-media backfill.
- Replacement of the well-graded lab-analyzed sand and gravel with a less expensive nongraded mix such as bank-run or pit-run gravel.
- In some cases, elimination of the need for perforated pipe.

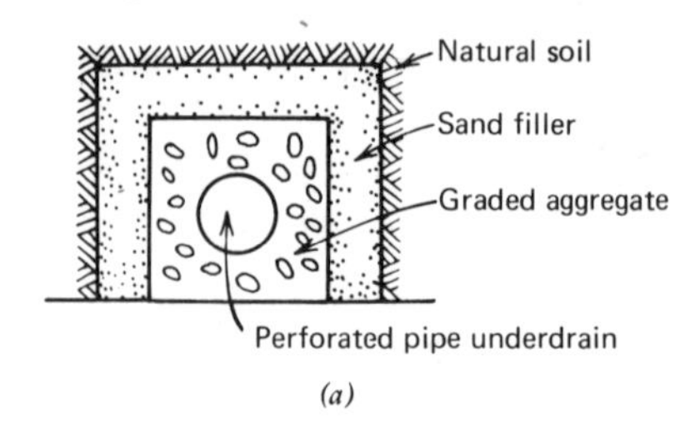

(a)

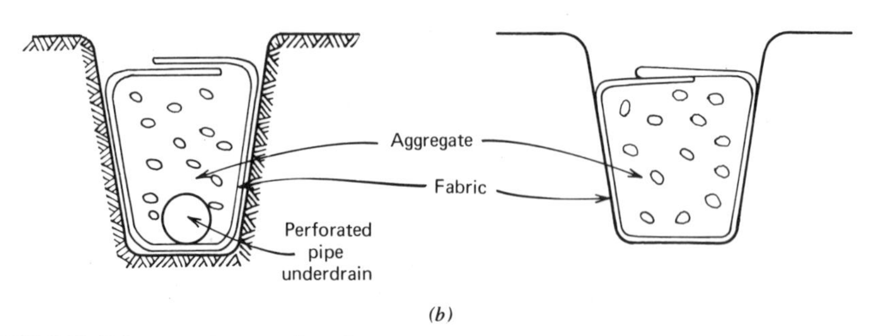

(b)

FIGURE 5.2. (*a*) Conventional pavement underdrain system compared to (*b*) two types of alternative underdrain systems using filter fabrics.

- Where only sand backfill is available, one can wrap the drainage pipe with fabric to act as a screening agent. The fabric thereby prevents the sand from entering the perforations in the pipe.
- Less trench excavation is required.
- Often, elimination of the need for trench shoring.

The possibility of entirely eliminating the need for the perforated pipe, that is, creating a French drain, is very attractive, and a few comments are in order about the typical design procedure.

The design of highway underdrains, that is, their size and the distance between outlets, is usually based on some form of solution of the Manning formula in the form of a nomograph. Cedegren[1] illustrates the procedure whereby a design infiltration rate is used along with the physical properties of pavement width and pipe gradient to arrive at a solution. Figure 5.3 can be used when one of the two unknowns (either pipe diameter or outlet spacing) is assumed to determine the other.

The substitution of a French drain system for a perforated pipe system, as in the lower portion of Figure 5.2, generally has sufficient void area to

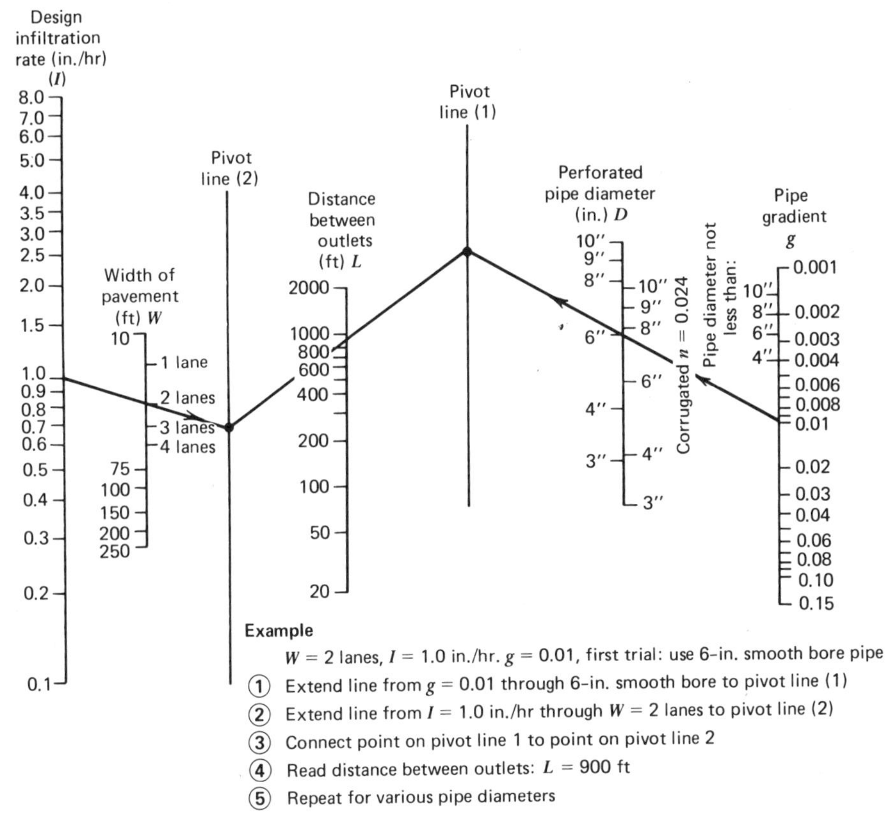

FIGURE 5.3. Nomograph for determining perforated pipe diameter and outlet spacing for pavement underdrain systems (after FHWA, Ref. 2).

remove the water (porosity of a poorly graded gravel is approximately 0.50), but requires the flow to take a very irregular path. This irregular and significantly longer flow path (along with the increased viscous drag forces that are mobilized) slows down the velocity of the water. Thus sedimentation of suspended particles can easily occur. With time this sedimentation could render the underdrain ineffective.

Two major considerations are required in making the decision of whether or not to use a French drain system. One concerns the slope (which must be greater than when using a pipe underdrain), the other the choice of filter fabric encapsulating the stone aggregate.

A number of case histories can best illustrate the use of fabrics in conjunction with pipe underdrain and French drain systems.

Case Histories

Benson[3] has reported that the Illinois Department of Transportation has used fabrics in connection with highway underdrains with very satisfactory results. General schemes are shown in Figure 5.4. Following are three separate case histories where fabrics have been used in connection with underdrain systems. In each case it should be noted that the fabric is also being used as a separator between the *in situ* soil and the stone filter material.

CASE 1. INTERSTATE 64 IN ST. CLAIR COUNTY. In a cut section of a proposed roadway that was to be placed on a high-water-content sandy

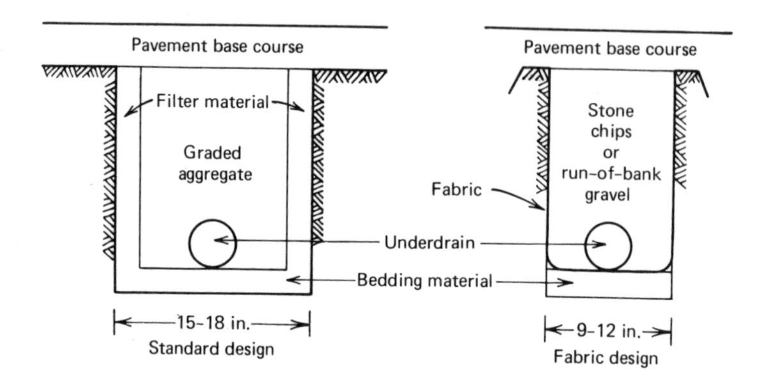

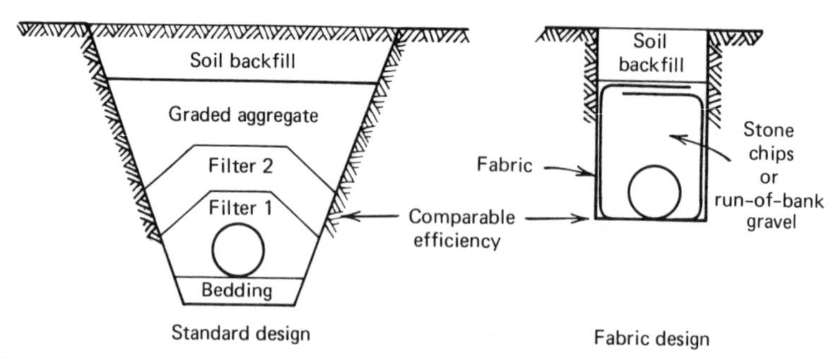

FIGURE 5.4. Various possible cross sections of underdrain schemes using fabrics.

silty clay the contractor elected to leave the road grade 2 feet higher than final elevation for greater stability during construciton. This presented a problem for underdrain trench stability, which was now 2 feet deeper than originally planned.

The plans called for a 10-inch perforated corrugated metal pipe with a coarse open-graded stone around the pipe and Mirafi 140 fabric around the entire system. However, during laying of the 20-foot pipe sections the sides of the excavation lost their stability and caved in before the pipes could be joined and properly aligned. It was decided to delete the pipe entirely because of high gradients at this particular location. The resulting fabric-wrapped stone drain has been performing excellently since completion. There is a steady flow of water, with no loss of fines, and the roadway is in excellent condition.[3]

CASE 2. INTERSTATE 24 IN JOHNSON COUNTY. In a related problem to the above, the subgrade remained unstable long after drainage ditches had been excavated. The soil was a high-water-content silt.

The district elected to use a system of transverse drains to dewater the silty soil, the seepage water being discharged into a median ditch collector. The silty nature of the soil meant that any drainage system involving an aggregate filter would almost certainly require a multielement system. As in the St. Clair County installation, the fabric was selected as one element of the filter; this time the fabric was Typar. The transverse drains employed underdrain pipe, but the main collector trench again had sufficient gradient (together with its cross section and highly permeable aggregate) so that the pipe could be left out. Again in this case, the fabric full of open-graded stone (coarse aggregate for concrete) provided the required draining capacity. This system, as the St. Clair application, continues to function very satisfactorily with no apparent loss of soil fines.[3]

CASE 3. INTERSTATE 55 IN CHICAGO. Another application of fabric underdrain utilization involved approximately 10 miles of reconstruction of the Stevenson Expressway (Interstate 55) in the Chicago metropolitan area. This stretch of roadway has for years had a history of water in the pavement structure, with obvious distress signs in both the pavement and shoulders. Drainage of the pavement structure appeared to be very important in any reconstruction plan since no internal drainage system had been built into the original improvement.

The severity of this problem was further complicated by the importance of the expressway and the very high daily traffic flow it accomodated. These factors required the design of a corrective solution that would remove the water from the pavement system as quickly as possible without complete pavement reconstruction, since this work was to be performed under traffic. The scheme selected consisted of a longitudinal drainage system installed by sawing through the existing stabilized shoulder at the joint between pavement and shoulder, and then trenching into the subgrade to the design depth. The drainage trench was then lined with Mirafi 140 fabric, then a 6-inch perforated bituminized fiber drain was placed in the bottom, and the remainder of the trench was filled with open-graded coarse aggregate. The fabric was then folded over the aggregate to form a complete envelope. In this situation, a pipe was considered necessary to provide proper drainage because of the very flat gradients involved. Performance of the system has been very satisfactory.[3]

CASE 4. AGRICULTURAL DRAINAGE. An interesting study combining theory, laboratory experiment, and field testing has been presented by Dierickx[4] on the potential use of fabrics to wrap agricultural drains or trenches. His results lead to the following conclusions:

- A thin fabric only acts as a substrate upon which a natural soil filter can be built up.
- A thick fabric markedly decreases the hydraulic gradient at the outflow surface.
- Fabrics with fine pores can be clogged with fine or colloidal soil particles, or iron and salt deposits.
- Organic filter materials, for example, peat, flax, and coconut fiber, are preferable (according to Dierickx) in agricultural drainage problems to fabrics on the basis of required thickness, high permeability, larger pores, and lower price.

Such results are of great interest in the selection of the appropriate fabric for a specific set of conditions. The last comment, however, is contrasted with other test data[5,6] on fabric-wrapped plastic drain tubes used in agricultural applications. These test results have given impetus to many field installations, some of which have the fabric placed around the pipe underdrain by the pipe manufacturer. Comparative tests are currently ongoing under the direction of ASTM Committees D-13.61 and F-17.65, with preliminary indication of little, if any, clogging, cake buildup, or dependence on fabric permeability in tests on fabric-wrapped drains.

CASE 5. HEALY AND LONG SYSTEM. A combined fabric interceptor and wrapped underdrain system was introduced by Healy and Long[7] in the early 1970s. A schematic diagram of their system is shown in Figure 5.5. Here it is seen that the vertical core (used for both support and water trans-

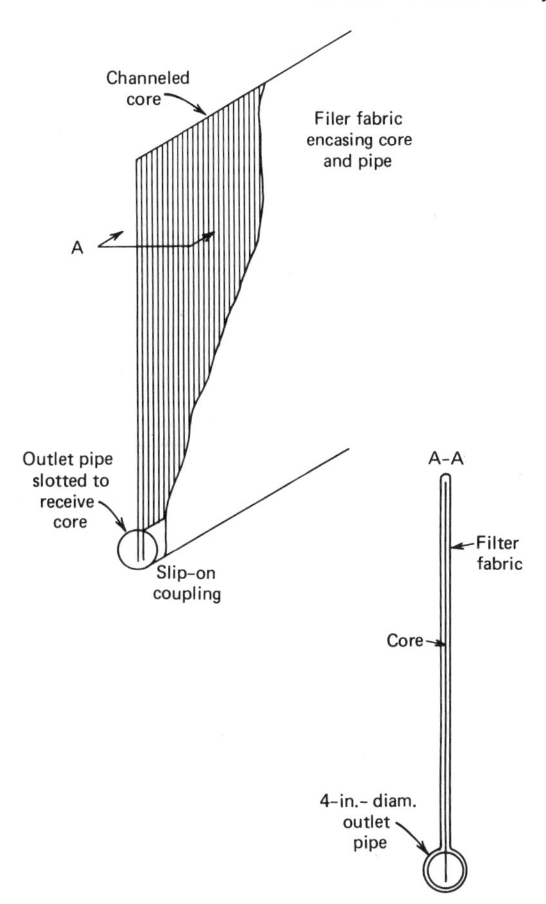

FIGURE 5.5. Components of prefabricated underdrain (from Ref. 8).

mission) intersects and actually penetrates the slotted underdrain pipe. The entire system is then wrapped with fabric. Three types of fabrics have been used; woven, nonwoven, and mat. Long and Healy[8] have recently developed design criteria based upon the characteristics of the *in situ* soil; namely, poorly graded, well-graded or nonhomogeneous (see Table 5.1 for details).

The method has been field tested, as seen in Table 5.2. Note the wide range in permeability of the soils where the installations were made. Of the six case histories cited, two were installed to stabilize slopes and four were installed to intercept ground water.

In their conclusion, Long and Healy feel that woven fabrics are the most suitable materials for this type of drainage application.

TABLE 5.1. General Requirements for Underdrain Fabrics

Type of Soil	Fabric Property	Type of Fabric		
		Woven	Nonwoven	Mat
Uniform	Opening size	$<D_{60}$	$D_{20} \leqslant$ O.S.[a,b] $\leqslant D_{60}$	Sieving test retains D_{50}
	Open area (%)	>5	>5	—
	Permeability[c]	—	—	$k_{mat} > 2k_{soil}$
Well-Graded	Opening size	$D_{30} <$ O.S. $< D_{85}$	$D_{30} <$ O.S. $< D_{85}$	Sieving test retains D_{85}
	Open area (%)	>5	>10	—
	Permeability[c]	—	—	$k_{mat} > 5k_{soil}$
Nonhomogeneous	Opening size[d]	$D_{20} <$ O.S. $< D_{80}$	$D_{20} <$ O.S. $< D_{80}$	Sieving test retains D_{80}
	Open area (%)	>10	>15	—
	Permeability	—	—	$k_{mat} > 2k_{coarse\ soil}$

Source: Ref. 8.
[a] O.S. = opening sizes.
[b] Between these sizes.
[c] Under design pressure.
[d] Based on finest soil.

TABLE 5.2. Summary of Observations on Field Installations

Location of Underdrain Installation	Date Installed	Type of Fabric	Soil Type	Permeability of Disturbed Samples (cm/sec)
Tennis court slope	June 1969 June 1970	Butterfly chiffon	Sand, silt, and clay	7.6×10^{-7}
Route 44-A	August 1970	Chiffon	Fracture rock to sandy silt	1.0×10^{-3}
Fellon Road	October 1970	Chiffon	Sandy silt	5.6×10^{-4}
Route 82, Haddam, Conn.	June 1971	Chiffon	Silty sand	1.0×10^{-4}
Retaining wall	August 1971	Chiffon	Clayey silt and sand	3.4×10^{-5}
Chaplin, Conn.	September 1974	Chiffon	Sandy gravel	2.0×10^{-2}

Source: Ref. 8.

5.2 Drainage in Earth Dams

Background

Combined with adequate strength to withstand the forces involved, proper seepage control in earth and earth/rock dams is of paramount importance. While the topic is adequately covered in standard references, such as Cedegren[9] and Sherard *et al.*,[10] an overview here is appropriate.

A homogeneous earth or earth/rock dam under steady-state seepage conditions will develop a zone of saturation, which will emerge on the downstream slope of the embankment, as shown in Figure 5.6(*a*). Such seepage will obviously cause erosion of the downstream slope, resulting in a subsequent loss of stability (loss of soil in the zone of passive earth pressure is occurring) and eventual failure of the structure. To simply use high-permeability soils that will correct the problem is not acceptable since reservoir losses will be too great for the supply.

The solution to this problem is to create a zoned embankment with a clay corewall and/or cutoff trench (to limit the seepage) and appropriate drains to remove what water does pass through the fine-grained soils. The function of such drains is to draw the zone of saturation away from the downstream surface of the embankment and safely within it (see Fig.

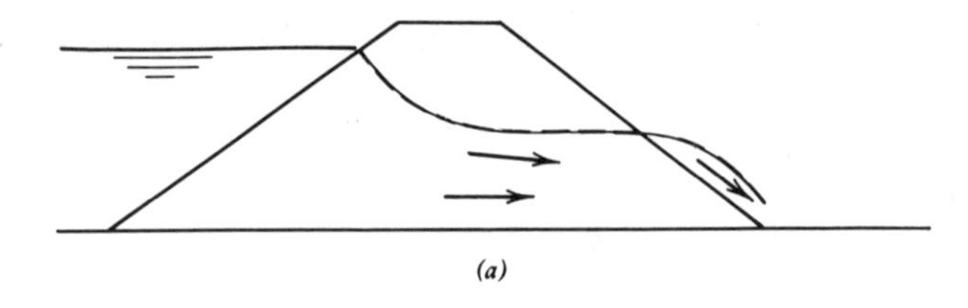

(a)

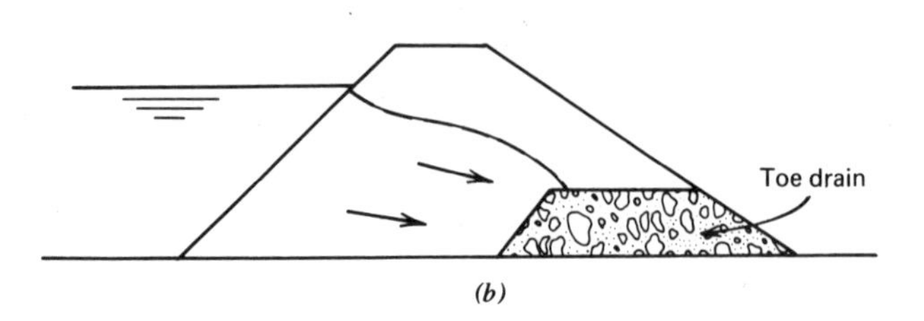

(b)

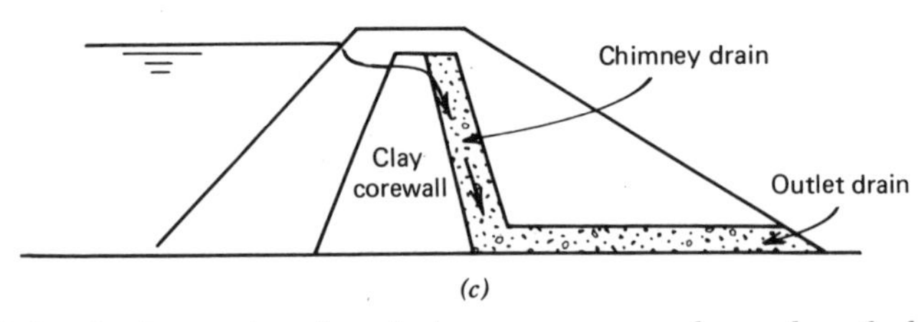

(c)

FIGURE 5.6. Drainage situations in homogeneous and zoned earth dams. Upper seepage path in (*a*) homogeneous earth dam; (*b*) homogeneous earth dam with toe drain; (*c*) zoned earth dam with chimney drain and outlet drain.

5.6(*b*)). Figure 5.6(*c*) shows a typical drainage system consisting of drainage galleries and chimney drains. Recall, however, that the situation is generally more complex than these sketches indicate (see Figure 3.1 for some actual cross sections).

Problems

The problems involved in the construction of toe drains, chimney drains, outlet drains, drainage galleries, and also transition zones are as follows:

- Most situations require carefully selected sands and gravels, which are scarce in many locations.

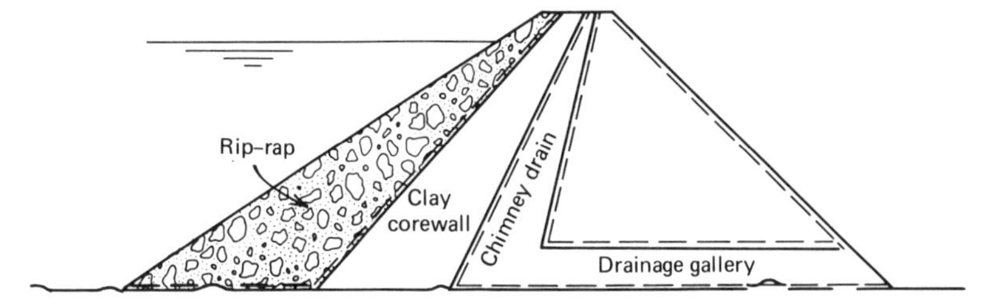

FIGURE 5.7. Dashed lines show various locations of fabrics used within zoned earth dams.

- Some situations require a filter system due to vastly dissimilar soils, for example, deformable clay in the corewall versus relatively non-deforming sand in the drainage system. Thus more than one material is required.
- Placement of such zones is very expensive, often resulting in the placement of more material than the contract calls for (at the cost to the contractor)—or endless arguments at the job site.
- The design itself is not analytically rigorous, and considerable judgement is required by the design engineer.

Fabrics are being tested and installed in earth dams for a variety of purposes connected with the solution of drainage problems. To be considered in such applications of fabrics is their permeability (both transverse and in-plane), clogging potential, strength, and longevity under the prevailing conditions.

A number of situations exist for the potential use of fabrics in the following cases, where earth dams using fabrics are being designed or constructed (see Figure 5.7).

Case Histories

The French have been particularly active in incorporating fabrics into various aspects of earth dam construction. These applications include:

- Collector drain replacement or protection to avoid use of gravel within the dam or at the downstream slope.[11,12]
- Seepage cutoff and drain collector.[12]
- Separation and drainage material under rip-rap protection of upstream slope.[11]
- Tensile reinforcement within the earth dam itself.[12]

CASE 1. Giroud *et al.*[11] report that in 1970 Bidim was used in the Valcros Dam in France as a drainage system (by encapsulating gravel) in the downstream toe and as a cover on the 3:1 upstream slope prior to placement of rip-rap. Samples were taken six years later and subsequently tested for tensile strength, which was somewhat reduced, and permeability, which remained the same as when originally placed.

CASE 2. Loudiere[12] reports on the use of fabrics in a number of different applications in earth dams. Figure 5.8 illustrates the various possibilities. In Figure 5.8(*a*) the fabric is used as a downstream drainage collector. In

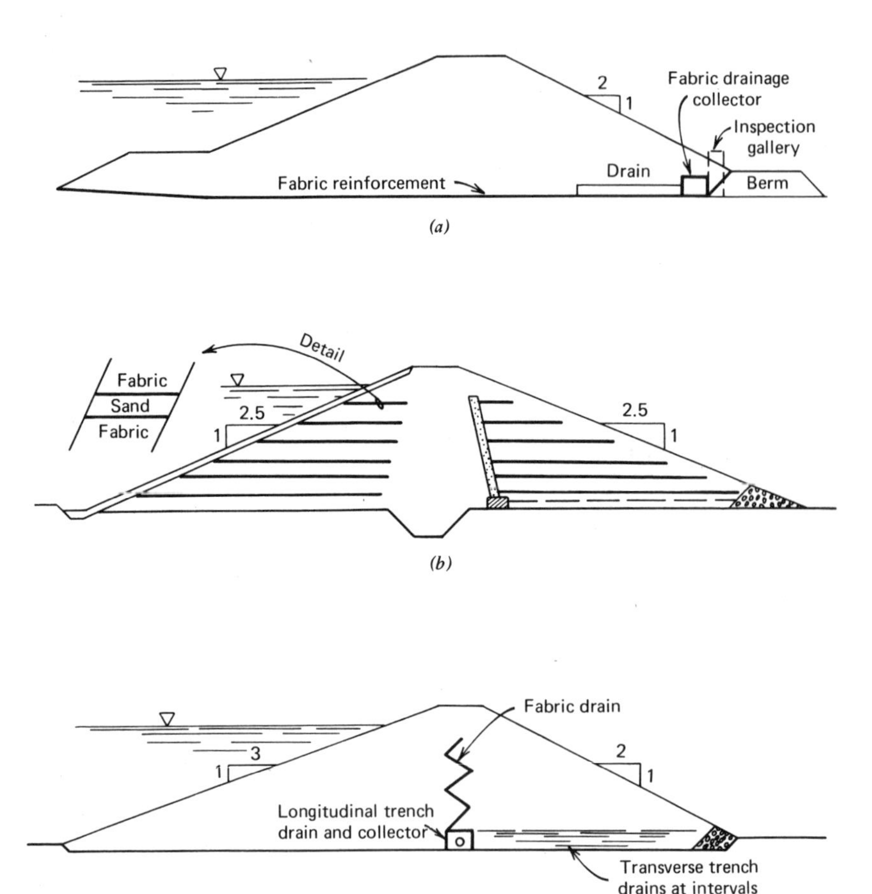

FIGURE 5.8. Possible uses of fabrics in earth dams as (*a*) downstream drainage collector; (*b*) tensile reinforcement; and (*c*) seepage cutoff and drainage collector (from Ref. 12).

Figure 5.8(*b*) horizontal layers of fabric are used as tensile reinforcement. Figure 5.8(*c*) illustrates the uses of the fabric first as seepage cutoff (i.e., similar to a chimney drain), then as a longitudinal drainage collector. Although not specifically mentioned, the transverse outlet system could also have incorporated fabric.

In these applications, which were earth dams on the order of 10 m in height, Bidim was used.

5.3 *Drainage Behind Retaining Walls*

Background

In Section 4.3 the use of fabric reinforcement in retaining walls is discussed. The design procedure that is presented assumes that the drainage system involved is adequate so that hydrostatic pressures are not acting along with the earth pressure. In this section we look more closely at this premise along with existing nonfabric and fabric drainage systems behind retaining walls.

The active earth pressure of soil acting on a permanent or temporary retaining wall is dependent on the unit weight of the soil, the height of the wall, and the coefficient of active earth pressure. This latter term depends on the shear strength of the backfill soil, and usually varies between 0.3 and 0.4. Figure 5.9(*a*) gives calculations for the backfill soil in a drained condition. The figure also gives the calculations if the backfill soil is not

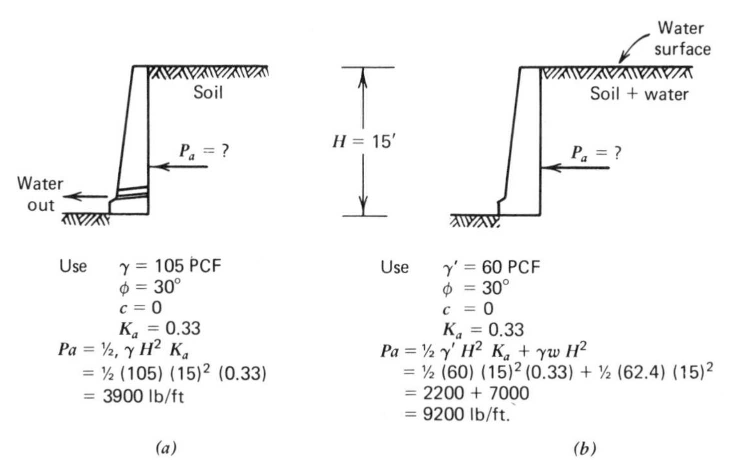

FIGURE 5.9. Earth pressure calculations showing the importance of adequately drained backfill soil for retaining wall design problems. (*a*) Soil drained; (*b*) soil not drained.

drained and hydrostatic pressures are allowed to build up, i.e., Figure 5.9(*b*). In this case a major influence on the total pressure against the wall is due to the water itself. The net effect of the water and soil is an approximate doubling of the earth pressure over that of properly drained backfill soil.

Problem

The removal of water from the backfill zone behind retaining walls is generally accomplished by a zone of high-permeability granular soil placed directly behind the wall for a thickness of 12 to 24 in. This vertical drainage zone then feeds weep holes, which pass through the wall itself or through an underdrain system, as shown in Figure 5.10. While both methods are effective, they are not without their share of construction problems and, to be sure, are expensive and a general "bother" to the contractor. Specifically:

- The sand backfill is a specified material and is usually purchased and transported from off site.
- The quantities of sand involved are generally small (12-inch thickness is specified in many situations); therefore, job loss of the material is high.
- Placement is difficult, with metal plates often forming the separation between the sand and the backfill—these plates being lifted as backfilling progresses.
- Overruns of placed sand are not uncommon, with the cost of excess material being borne by the contractor.
- Weep holes are troublesome to construct and result in holes in formwork.

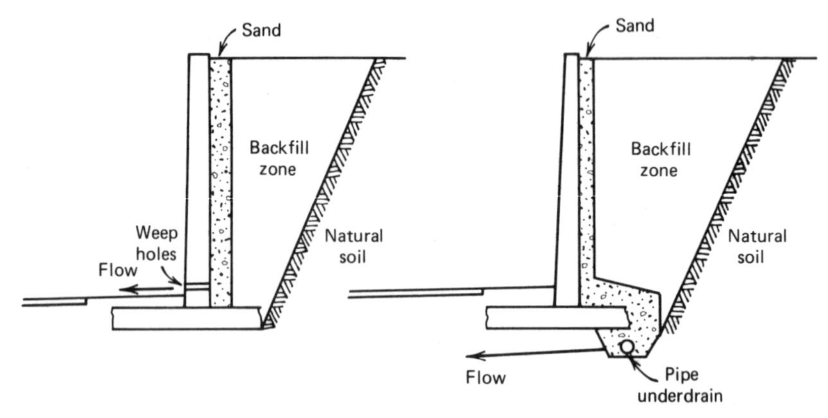

FIGURE 5.10. Standard methods of draining the backfill zone behind retaining walls.

- Often the weep holes are not clear after stripping the forms, requiring the hardened concrete to be chipped out of them.
- Installation of underdrains holds up construction of the wall itself.
- Backfill of the underdrains is very important since the footing will be founded (at least in part) on this area.

Needless to say, the use of construction fabrics to replace or alleviate some of these problems of constructing an adequate and long-term functioning drainage system behind retaining walls is of great significance. In the use of fabrics for this application it is important to realize that the in-plane permeability of the fabric is being relied upon. Furthermore, this in-plane permeability must function while the fabric is being subjected to the normal pressure of the soil adjacent to it.

Case Histories

CASE 1. In an interesting review paper, McGuffey[13] has described many of the fabric uses utilized by the New York State Department of Transportation. One novel use has been to line the sloped ground surface between the original soil and the backfill behind concrete retaining walls. See Figure 5.11(*a*) for McGuffey's approach or Figures 5.11(*b*) and 5.11(*c*) for our alternative approaches. All schemes are intended to eliminate hydrostatic pressure buildup behind retaining walls, which are usually designed based upon active earth pressures. Once the horizontally flowing water is intercepted, it must drain vertically downward to an underdrain system or to weep holes through the bottom of the stem section of the wall.

The possibility exists that in some cases the sand backfill might be completely eliminated and the fabric placed directly against the back of the concrete wall, as also shown in Figure 5.11(*c*). Provided that sufficient in-plane drainage capacity were available to transport the water, this application could represent a tremendous saving in the labor-intensive item of drainage layers behind retaining walls.

CASE 2. Nicolon fabric has been used in a similar manner but behind sheet pile walls,[14] where it not only allows for the release of ground water, but also prevents fines from escaping from the backfill soil through the joints in the wall (see Figure 5.12). This type of application could be used in all types of temporary or flexible wall systems consisting of steel sheet piles, soldier beams and lagging, concrete or wood tongue and groove panels, or even skeleton sheeting.

CASE 3. Two commercially available fabrics that have little or no resistance to in-plane water flow are Enkadrain (see Section 2.5.6) and

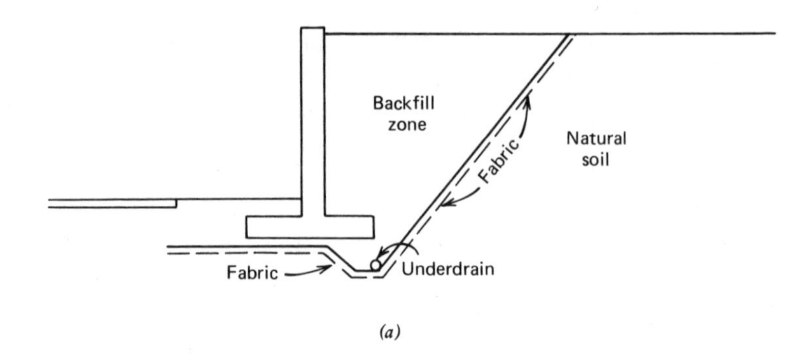

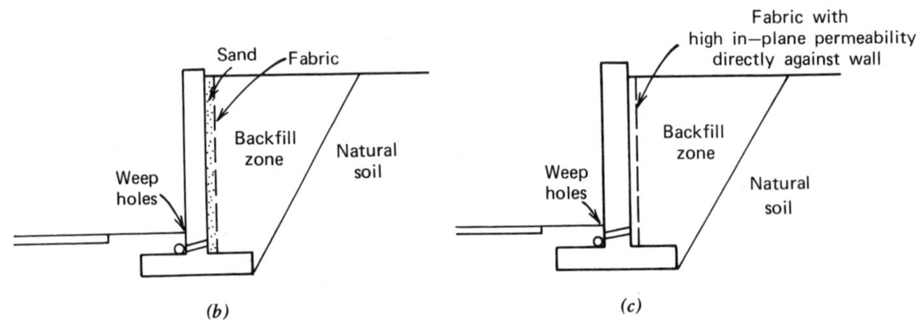

FIGURE 5.11. Possible schemes for fabric use behind retaining walls (*a*) McGuffey's approach (Ref. 11); (*b*), (*c*) alternative approaches.

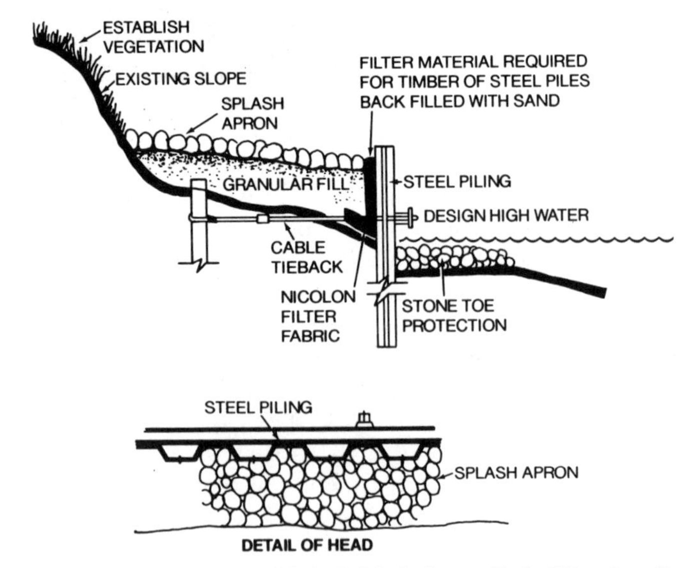

FIGURE 5.12. Use of fabric behind sheet pile bulkhead walls.

Filtram (see Section 2.5.24). Both of these materials are composites made by sandwiching a material with an open network between one or two sheets of filter fabric. The resulting composite is then placed, or suitably attached, against the retaining wall or other structure to be drained. The filter fabric prevents the soil from moving while allowing the water to pass through it and into the inner material. The water is then rapidly drained from the system.

The major caution that must be exercised is to be sure that the lateral earth pressures from the soil being retained do not collapse the inner high-permeability material, thereby closing off the flow channels while it is in service.

This particular application of extremely high in-plane permeability fabric composites should have a marked influence on backfilling procedures behind many different types of retaining structures. It appears to represent a definite technical improvement over current practices at a decidedly lower cost to both contractor and owner.

5.4 *Fabric Drains to Accelerate Settlement*

Background

Post-construction settlement of structures in saturated fine-grained soils is so significant that a major portion of geotechnical engineering is devoted to the topic. The classic work of Terzaghi in analyzing the problem (both the amount of settlement and the time for settlement to occur) marks the very beginning of soil mechanics as a separate field of study in civil engineering.[15] While both aspects of consolidation settlement, the actual amount and the time, are important it is the latter that often gives rise to long-term problems. The time for consolidation can be predicted using the following equation:

$$t = \frac{H_v^2 T_v}{c_v} \tag{5.1}$$

where t = time for settlement to occur
H_v = drainage path length in vertical direction (often the depth of the stratum)
T_v = vertical time factor depending on the percentage of consolidation
c_v = coefficient of consolidation for vertical flow

It is the coefficient of consolidation (a laboratory-determined property of the soil in question) that can lead to intolerably long times for settlement to occur. Very low c_v values can lead to consolidation times of tens and hundreds of years. Equally important in many cases are thick layers of compressible soils which, being a squared term in the equation, also lead to long consolidation times.

In order to hasten consolidation time periods a combination of sand drains and surcharging has been used since the 1930s. Sand drains are vertical columns of high-permeability sand or gravel that are installed continuously through the fine-grained subsoil that is causing the problem. A preconstruction fill, called a surcharge loading, is then placed over the entire site covering the sand drains. This surcharge mobilizes excess pore pressures in the soil water, which then flows horizontally to the sand drain, and then vertically within the sand drain to eventually escape through a sand blanket or other permeable horizontal stratum. By so doing the equation for the time for consolidation becomes

$$t = \frac{H_h^2 T_h}{c_h} \tag{5.2}$$

where t = time for settlement to occur
H_h = drainage path length in horizontal direction (half the distance between sand drains)
T_h = horizontal time factor depending on the percentage of consolidation, the sand drain diameter and the sand drain spacing[16]
c_h = coefficient of consolidation for horizontal flow

It is easily seen by comparing equations 5.1 and 5.2 that the time for consolidation settlement to occur is drastically reduced when using sand drains. This comes about by c_h values often being larger than c_v values, due to horizontal stratification, and shorter distances for flow in the horizontal versus the vertical directions. Note should be made that the *amount* of settlement is not reduced, only the *time* for this settlement to occur. Upon achieving the settlement desired, the surcharge, or a portion of it, is removed, and the final structure is built.

Construction of Sand Drains

The standard method of constructing sand drains is by the installation of a sand-filled steel pipe (called a mandrel) through the compressible stratum. The mandrel is then withdrawn, leaving a column of sand behind. This process is repeated at various spacings throughout the compressible soil. Typical sand drain diameters are 6 to 24 inches, with 5 to 20-foot spacings being common. An alternative method is to place the sand drains by using a continuous flight auger with the sand being introduced through a central core pipe as the entire unit is being withdrawn.

With either installation method, two fundamental problems exist:

- The sand clogs at the sand drain/soil interface, thereby restricting water from entering the sand drain.
- The sand drain offers no resistance to lateral shifting of the soil as nonuniform settlements or lateral shear strains occur.

The above problems can be avoided entirely, or at least partially offset, by using fabrics to encapsulate the sand or using fabrics by themselves as the complete drain. The following case histories illustrate these two situations.

Case Histories

CASE 1. The Chiyoda Chemical Engineering and Construction Co., Ltd.[17] of Japan has a patented method for constructing fabric-enclosed sand drains. Called the Chiyoda Pack Drain Method®, it consists of a polyethylene monofilament tube of varying diameter running the length of the compressible stratum. This net-type tube with 1.2 by 1.6 mm openings is inserted into a crane-mounted casing that drives the assembly into position for sand filling. The casing is withdrawn by a crane while the fabric-filled sand column is left behind with the aid of applied air pressure. The construction sequence is shown in Figure 5.13.

As with other applications of fabrics in drainage, the proper fabric can be considered to be a part of the drainage system, not merely used as a separating medium. Therefore, it is quite likely that smaller diameter sand drains can be utilized in comparison to standard procedures, but the design process is not currently documented or field tested.

In the United States, a similar system under the trademark Fabridrain™ is available under license from Construction Techniques, Inc., Cleveland, Ohio.

CASE 2. An extension of the preceding method begs the question, "Why use the sand at all?" Many fabrics are known to possess excellent in-plane permeability which, if properly placed, could transmit large quantities of water from a saturated consolidating soil stratum to an off-site location. The concept was first tried in 1948, not with fabrics, but with cardboard wicks. Kjellman[18] threaded these cardboard wicks throughout the compressible soil with excellent consolidation results. Partly because of the required construction equipment (reassembling a gigantic

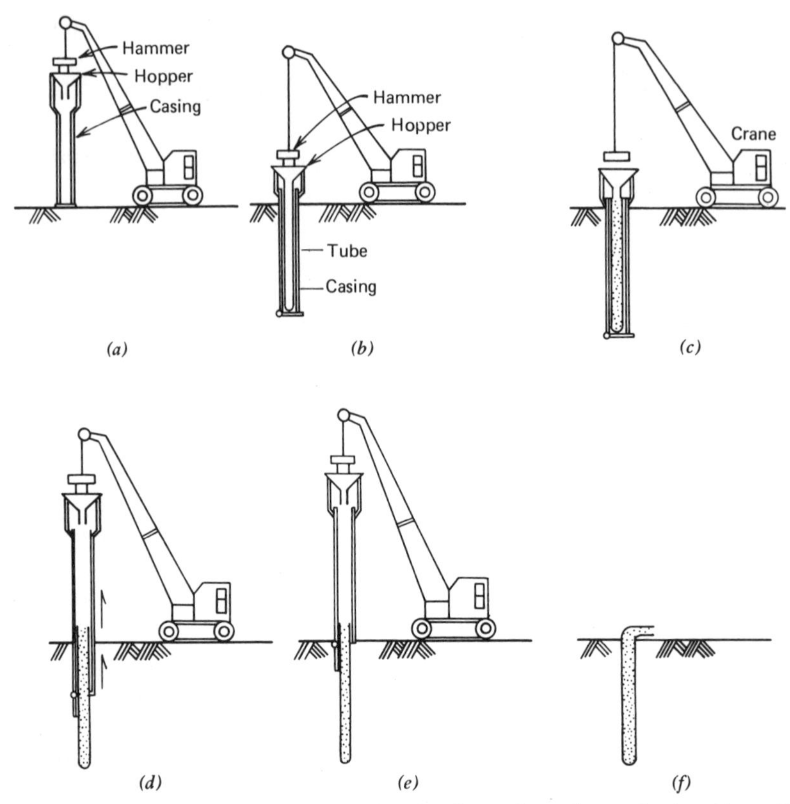

FIGURE 5.13. Schematic diagram of Chiyoda Pack Drain Method® of installing fabric-enclosed sand drains. (*a*) Casing operation starts; (*b*) polyethylene fabric tube is inserted into casing; (*c*) fabric tube is filled with sand; (*d*) casing is pulled out; (*e*) installation of Chiyoda Pack Drain is completed; (*f*) completed pack drain sand pile (from Ref. 17).

sewing machine), however, the technique has not been used to a large extent, at least in America.

Another variation, using fabrics, has been recently introduced. Risseeuw and van den Elzen[19] are using 30-centimeter-wide strips of nonwoven polyester Colbond KH630 fabric. As shown in Figure 5.14, installation of the continuous fabric strips is by means of a crane-mounted telescopic lance. The fabric, on a roller, is fed into the top of the lance where it is clamped in placed. The lance is then driven into the soil with the roller playing out the fabric as it penetrates. At the desired depth the fabric is released and the lance retrieved to move on to the next location. The in-place fabric is cut off at the ground surface. It could easily be connected to fabric placed horizon-

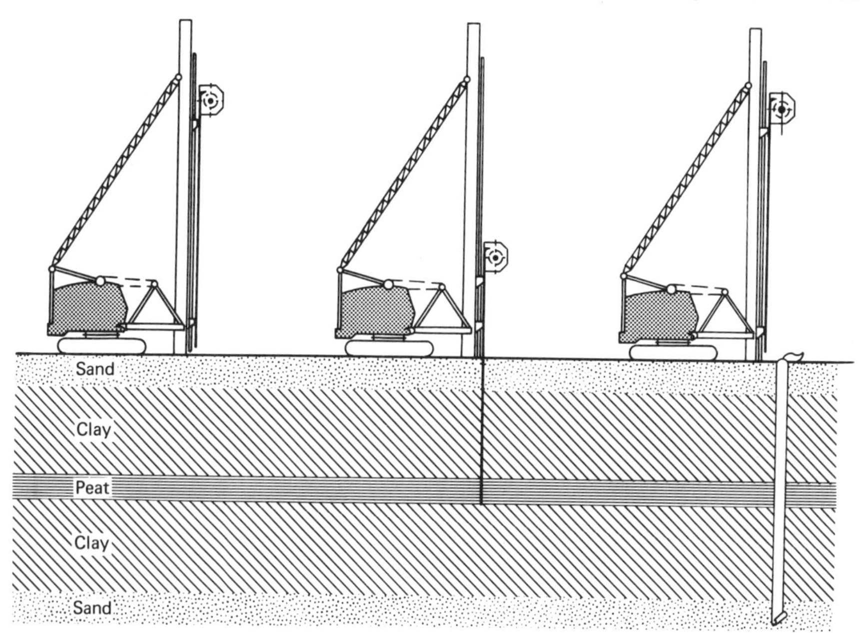

FIGURE 5.14. Installation stages for vertical drainage columns using fabrics (from Ref. 19).

tally on the ground surface to conduct the water out of the surcharge fill area, thereby eliminating the need for a sand blanket.

Field testing of this technique is well advanced, with a number of case histories being reported in reference 19, the largest being the installation of 180,000 meters of fabric drain. Results are comparable to the behavior of soils consolidated using standard sand drains. Note should be made that in penetrating hard soil, or strata of hard soil, jetting will be required to advance the lance containing the fabric. Thus considerable remolding will probably occur, offsetting any potentially available horizontal stratification. No comments regarding the nature and extent of the smear zone common to sand drain installation via mandrel driven or augered types are known.

5.5 References

1. H. R. Cedegren, *Drainage of Highway and Airfield Pavements*, Wiley, 1974.
2. *Guidelines for the Design of Subsurface Drainage Systems for Highway Structural Sections*, Federal Highway Administration, Washington, D.C., 1973.

3. G. R. Benson, "Filter Cloth in Illinois," *Highway Focus*, Vol. 9, No. 1, May 1977, pp. 43-48.
4. W. Dierickx, "The Influence of Filter Materials and Their Use as Wrapping Around Agricultural Drains," *C. R. Coll. Int. Sols Text.*, 1977, Vol. II, pp. 225-229 (in French).
5. L. C. Benz, E. J. Doering, G. A. Reichman, and R. F. Fullett, "Evaluation of Some Subsurface Drainage Envelopes," *Proceedings of the Third National Drainage Symposium*, American Society of Agricultural Engineers, St. Joseph, Mich., 1976; R. S. Broughton, B. English, S. Ami Damont, E. McKyes, and J. Brosseur "Tests of Filter Materials for Plastic Drain Tubes," *ibid.*
6. L. S. Willardson and R. E. Walker, "Synthetic Drain Envelope/Soil Interactions," unpublished.
7. K. A. Healy and R. P. Long, "Prefabricated Subsurface Drains," *Highway Research Rec.*, No. 360, 1971, p. 57.
8. R. Long and K. Healy, "Fabric Filters on Pre-Fabricated Underdrains," *C. R. Coll. Int. Sols Text.*, 1977, Vol. II, pp. 237-241.
9. H. R. Cedegren, *Seepage Drainage and Flow Nets*, Wiley, New York, 1967.
10. J. L. Sherard, R. J. Woodward, S. F. Gizienski, and W. A. Clevenger, *Earth and Earth-Rock Dams*, Wiley, New York, 1963.
11. J. P. Giroud, J. P. Gourc, P. Bally, and P. Delmas, "Behavior of Nonwoven Fabric in an Earth Dam," *C. R. Coll. Int. Sols Text.*, 1977, Vol. II, pp. 213-218 (in French).
12. D. Loudiere, "The Use of Synthetic Fabrics in Earth Dams," *C. R. Coll. Int. Sols Text.*, 1977, Vol. II, pp. 219-223 (in French).
13. V. C. McGuffey, "Filter Fabrics for Highway Construction," *Highway Focus*, Vol. 10, No. 2, May 1978, pp. 1-21,
14. Product Literature, Nicolon Corporation, U. S. Textures Sales Corp., 4229 Jeffrey Dr., Baton Rouge, La. 70816.
15. K. Terzaghi, *Erdbaumeknic*, Deuticke, Leipzig, 1925; also, *Theoretical Soil Mechanics*, Wiley, New York, 1942.
16. R. A. Barron, "Consolidation of Fine-Grained Soils by Drain Wells," *Trans. ASCE*, Vol. 113, 1948, pp. 718-754.
17. Product Literature, Chiyoda International Corp., 1300 Park Place Building, 1200 6th Ave., Seattle, Wash. 98101.
18. W. Kjellman, "Accelerating Consolidation of Fine Grained Soils by Means of Cardboard Wicks," in Proceedings of the Second International Conference on Soil Mechanics and Foundation Engineering, Rottendam, 1948.
19. P. Risseeuw and L. W. A. van den Elzen, "Construction on Compressible Saturated Subsoils with the Use of Nonwoven Strips as Vertical Drains," *C. R. Coll. Int. Sols Text.*, 1977, Vol II, pp. 265-271.

6

Fabric Use in Erosion Prevention

Erosion prevention and/or control is big business! With soaring real estate prices, accelerating building in coastal areas, and the ever attractive magnetism of water for recreational purposes this business will present ever increasing challenges. Fabrics can, and have, played an important role in meeting this challenge, and this section describes some of the current fabric applications.

But the forces involved in water and wind erosion are enormous. The beauty unleased by a violent storm over a lake or ocean front can only be matched by the destruction to natural and synthetic objects in its path—failures abound. While this section emphasizes the positive uses of fabrics, failures of fabric erosion control systems have occurred. For example, Miller[1] describes in some detail the failure of a stone rip-rap revetment underlain by filter fabric. It should be kept in mind that while designing an erosion prevention system using fabrics there is currently an absence of quantitative design information in the following areas:

- Admissible uplift or suction forces.
- Resistance against wave action.
- Sliding against soils to be protected.
- Water movement into and out of the soil mass being protected.
- General system permeability.

If an improper choice of fabric or system is made in any of these areas, failure could result.

6.1 Fabric as a Soil-Retaining Mechanism

Background

Erosion of exposed, unprotected soil is a problem that demands solution. Due to increased construction and increased population density, the damaging results of erosion have carried great impact and have become very noticeable in our environmentally minded society (see Figure 6.1 for such effects).

The quantities of soil that water or wind can move are staggering. Over four billion tons of topsoil are eroded each year by rainfall alone.[2] The force of rainfall hitting the ground at 20 miles per hour can loosen and erode any bare soil. Additionally, the rainfall compacts the upper portion of the soil, thereby reducing its absorptive capacity, creating more runoff, which results in sheet, rill, and gully type erosion.

FIGURE 6.1. Typical erosion pattern of bare soil.

Thus water moving over a soil surface usually results in two undesirable events:

1. Damage to the soil surface via erosion.
2. Deposition of the eroded soil in adjacent areas (commonly a nearby waterway or sewer system).

To protect against such erosion two courses of action can be taken. First, use natural vegetation, for example, grasses, weeds, sod, and so on. The second is to protect the slope by some type of paving. Standard types of pavement are concrete, asphalt, rip-rap, gabions, and others.

The use of natural vegetation obviously assumes that such vegetation can grow on the slopes in question. Critical in this regard is reduction of the velocity of the runoff water since this is the key parameter in soil loss. For example, the universal soil loss equation is as follows[3]:

$$A = RKLSCP \tag{6.1}$$

where A = soil erosion loss per unit area
R = rainfall factor
K = soil erodability factor
L = slope length factor
S = slope gradient factor
C = cropping-management factor
P = erosion control practice factor

In the above equation K, L, and S are interdependent with the runoff water velocity.

It seems natural to use fabrics in this type of application, although there are numerous competing systems.[2]

Case Histories

CASE 1. The Cold Regions Laboratory in the Corps of Engineers investigated two types of erosion control fabrics in reducing soil loss at construction sites in cold regions.[4] The two fabrics were (1) a white, nonwoven fiberglass matting produced by Owens Corning Fiberglas Corporation and (2) a plastic netting interwoven with $\frac{1}{4}$ inch strips of brown paper produced by the Gulf States Paper Company.

Five test plots each measuring 8 by 46 feet were constructed on a 25° north facing slope, which consisted of a mixture of varved clays and highly erodable fine sandy loam. The fabrics were installed on plots previously treated with fertilizer (15-15-15) and grass seed (Vermont Soil Conservation mixture). The fiberglass fabric was installed on three test plots, the paper fabric on one test plot, and the remaining plot was left bare of seed, fertilizer and fabric and used as a control.

Soil loss measurements after one year indicate that both fabrics are highly effective in reducing soil loss. Both fabrics had an effectiveness of about 96 percent when compared to the control plot. The fiberglass plots had an average soil loss of 2.5 tons per acre and the paper fabric had a soil loss of 2.7 tons per acre compared to the control plot, which lost 76.0 tons per acre.

CASE 2. A recent summary note[5] discusses the acceptance of synthetic mesh nets and perforated sheets for erosion control. The fabric materials provide a cover to protect the soil base from direct water flow, yet still allow seeded vegetation to germinate and grow. The more biodegradeable materials, like jute matting and asphalt tack coats, can break down and leave the slope unprotected. Fabric mats are persistent synthetics that remain on the ground and continue to give protection for longer periods.

One product offered uses a layer of degradeable fiberous material connected to a woven synthetic fiber fabric. The paperlike material offers protection to the underlying soil and at the same time allows the seeds to grow with no problem.

In most cases, materials that allow relief for hydrostatic uplift pressures are needed. This can be a severe problem where the level of subsurface water changes quickly and suddenly. In such cases, synthetic filter fabrics work well. Staked in place, they allow water, but not soil particles, to pass through without fabric displacement.

Fabrics can be of either woven or nonwoven; however, the woven fabrics usually have larger pore sizes than nonwovens and are used in soils of large particle size.

Nonwoven fabrics are used where fine soils, such as clay or silt, are found. Where hydrostatic uplift is expected, these fabrics must be of sufficiently high permeability so as not to present a problem. The fabric's voids may be blocked by the fine soil. The resulting pressure may then force the fabric to balloon out or tear away from the area being protected, allowing soil loss.

Care should be taken to choose the right size of fabric opening, that is, EOS or percent open area, for each case. Underlying soil samples should be taken and analyzed before specifying the filter fabric.

In this particular application where the fabrics are exposed to the atmosphere during their useful life, it is important to recognize that ultraviolet degradation will reduce the long-term strength of the fabric to varying degrees. Many fabrics have inhibitors added to retard degradation and, indeed, some fabrics are less susceptible than others. Care in fabric selection vis-à-vis the application and environment is necessary.

6.2 Fabric Liners for Erosion Control Systems

Background

In most situations where continuous flowing water is involved fabric protection of the underlying soil in itself is not sufficient. The currents involved will eventually undercut the soil beneath the fabric, moving the fabric out of position, rendering the soil base exposed and thus susceptible to erosion. In themselves, fabrics are difficult to submerge since the density is less than that of water. A number of different failure mechanisms are possible to occur.

- Attack at the toe of an underwater slope, leading to bank failure and erosion. This normally occurs in falling rivers at medium stage or lower.
- Erosion of soil, caused by current action, along the bank.
- Sloughing of saturated, cohesive banks which are incapable of free drainage and fail during rapid drawdown.
- Flow slides (or liquefaction) in saturated silty or sandy soils.
- Erosion of the soil by seepage out of the bank during low flows.
- Erosion of the upper bank, river bottom, or both due to wave action from wind or passing boats.

In such instances some type of erosion control system is usually placed on top of the soil. The most common types are[6]

- Stone rip-rap.
- Concrete pavement.
- Articulated concrete mattresses.
- Transverse dikes.
- Fences (transverse to stream flow).
- Asphalt pavement.

- Jacks (wood, concrete, or metal rods in the shape of a toy jack).
- Vegetation.
- Gabions (stone-filled wire cages).
- Bulkheads.

Each of the above methods is described in Ref. 6. Considerable literature on extended studies is available as well.

In most of these installations a granular filter is required on top of the soil and must be protected and beneath the structure—for example, rip-rap, gabions, or mattresses. This granular filter provides a number of functions: it retards loss of soil, eliminates hydrostatic pressure, and provides bedding for underlying materials. As necessary as the granular filter is in the adequate performance of an erosion control system, it is expensive to obtain and to place, thus suggesting an alternative approach, for example, fabrics.

Fabric Solution

Construction fabrics have been used as a substitute for granular filters beneath erosion control structures (or to augment them) in a number of different situations. To be sure, it is a demanding use often requiring of the fabric some, or all, of the following properties.

- High permeability.
- High friction resistance with soil.
- High tensile strength.
- High elongation.
- Good resistance to tear.
- Good resistance to puncture.
- Long life.
- Resistance to uv degradation.
- Dimensional stability.

When these demands are combined with harsh environments such as beaches, open sea water, high velocity rivers, and so on, both caution and care must be exercised. The results of an unsuccessful installation can often be seen within a short period of time. Thus this is somewhat more of a challenge than are the fabric uses described in earlier sections, for example, reinforcement or drainage, where the construction fabric is buried and not available for constant scrutiny.

Fabric Installation

The performance of fabrics placed as liners for erosion control systems is largely based on the geometry, location, local conditions, and fabric type, but a general guideline can be suggested.

- It is very advantageous to remove the water from the immediate vicinity of fabric placement if at all possible. This can be done by a small cofferdam (see Section 8.3), diverting the water, dewatering, or by working at low flow or low tide. The result will be a more productive effort and a better job.
- Prepare the slope upon which the fabric is to be laid as much as possible. At the minimum remove debris, rocks, stumps, and so on, fill in large voids or slope irregularities.
- Compact the slope, if possible. Even hand rolling is helpful in minimizing differential movement of the completed erosion control structure.
- If working in the dry, excavate the toe trench at the intended lower elevation of the fabric. Then, spread the fabric out on the slope and place and anchor it into the toe trench. Anchorage can be made by pins or stakes, or by placing soil or rock on top of the fabric, see Figure 4.7, for example. Proceed up the slope in this manner.
- If working in the wet, the toe trench must be excavated and simultaneously the fabric must be placed in it and anchored. Thus you will be working in two directions, both along the water's surface and up the slope as well. It is important that the fabric is anchored to the slope along its entire area under the water, since water behind the fabric can lift it out of place and wave action could tear it, or rip it out of the toe trench. In cold climates, ice formation on the fabric can be very troublesome.
- The amount of fabric overlap depends on the nature of the *in situ*, slope soil (mainly its compressibility), on the type of erosion control structure to be placed on the fabric (mainly its weight), and on the deformation characteristics of the fabric itself (mainly its creep behavior). Lap distances of 2 to 5 feet appear to be the customary range of values. If the fabric ends are to be joined by bonding or sewing, the joint strength must be equal to or greater than the strength of the fabric itself.
- Secure the upper end of the fabric in a trench or other suitable anchorage. Surface water should not be able to get beneath it and it

should be protected against vandalism or prolonged exposure. Keep a minimum amount of the fabric exposed at all times, particularly in warm, sunny climates when using fabrics made from uv-sensitive fabrics.

- If heavy materials are to be placed on the fabric, for example, stone rip-rap, protect it from puncture with a 3 to 5-inch layer of sand or run-of-bank gravel. Start at the toe and work up the slope with this protective layer and with the final erosion control structure.

Case Histories

CASE 1. The Maine Department of Transportation was faced with the continual erosion of a 35-foot-high highway embankment of Route 118 in Waterford, Maine, adjacent to the Crooked River.[7] The *in situ* soil was a very clean medium sand which easily erodes during high water flows. When erosion occurred, the upper embankment fell into this toe area in an attempt to reestablish equilibrium, leaving a scarp that required continual maintenance.

It was decided to rip-rap the lower portion of the slope to an elevation above the high water level. However, the migration of the clean sand through the rock boulders was a problem. Fabric was chosen to act as a separator between the sand and the rock boulders and as a filter to allow water to migrate in either direction, depending on the water's elevation. Head differences force water from the sand to the boulders during times of low river levels, this action reverses during times of high river levels.

The fabric selected by the contractor was Bidim C-22, which was toed in at both the bottom (into the river bed itself) and at the top (along the slope). Results are apparently satisfactory.

CASE 2. Since 1967, 2 million square meters of submerged river bank along the Yssel River in the Netherlands have been protected using fabrics.[8] The current solution evolved from earlier trials using sandbags in 1957 to nylon double-fabric filled with sand in 1960. Further development led to the following procedure, which has been used to a wide extent.

A porous construction fabric is stretched over a single layer of willow poles at large spacing for flotation purposes. A number of fascine poles are laid perpendicular to each other at one-meter spacings on top of the fabric and tied at their crossings. The fascine poles are made out of small pieces of willow. The "raft" is now moved out into the water in the upstream direction to its intended location. The flow of water is used to press the raft to the bottom. It is sometimes partly loaded with stone, depending on the water velocity and depth. After the mattress is completely lowered onto the slope to be protected it is covered with stone rip-rap. Then the connection to the sinking beam is broken, floats to the surface, and is reused for the next unit to be built.

Regarding the type of fabric, Hoogendoorn[8] feels that for sandy riverbanks monofilament fabrics are preferred. Experience has also led to the use of fabric openings on the order of 340 to 850 microns. The technique is patented.

CASE 3. Barrett[9] describes a number of case histories on the use of fabrics as liners for different coastal structures. A rubble revetment in Deerfield Beach, Florida in 1962 is discussed. The bank slopes were approximately 2:1 and the construction fabric was spread with rocks placed directly on it. The weights of the rocks were from 500 to 5000 pounds. In four years there was no settling of the revetment, nor was any maintenance required.

There have been many other revetments constructed where rocks of similar weights have been placed directly on the filter fabric. However, it is Barrett's opinion that a layer of gravel or crushed stone should be placed immediately on top of the filter fabric. Since this gravel layer would not be called upon to perform a filter function, whatever material is locally available and is most economical may be specified. The purpose of this layer is to act as a pad to prevent puncture of the fabric by the heavier rocks when movement occurs during a storm or hurricane. The size of the structure, the armor stones, and the intermediate stones will naturally dictate what size material should be used in this protective pad. The stone in this pad should be large enough so that it cannot pass through the rock layer above it. In various revetments the size of the material used in this protective layer has varied from $\frac{3}{8}$ inch gravel to one-man stone (100 to 150 pounds).

CASE 4. Dunham and Barrett[10] describe a series of projects that use fabrics as liners for seawalls.

Figure 6.2 is a cross section of a large stone seawall with a construction fabric built in 1970 for private development in Southern California. The contractor said that the only problem with the fabric was the tendency of the 2-foot (0.6-meter) overlaps to pull apart with the use of stone of this size. He used sheets mostly in 36-foot (11-meter) widths, laying them side by side along the slope, each with sufficient length to cover the entire section from toe to crown. Although skeptical of the fabric at first, he is now an ardent proponent of its use in all types of shorefront structures and keeps a small stock on hand for odd jobs and for emergency use.

The seawall shown in Figure 6.3 was constructed at the north end of Carolina Beach, North Carolina, in 1970 by the Corps of Engineers.

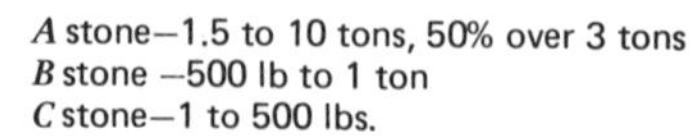

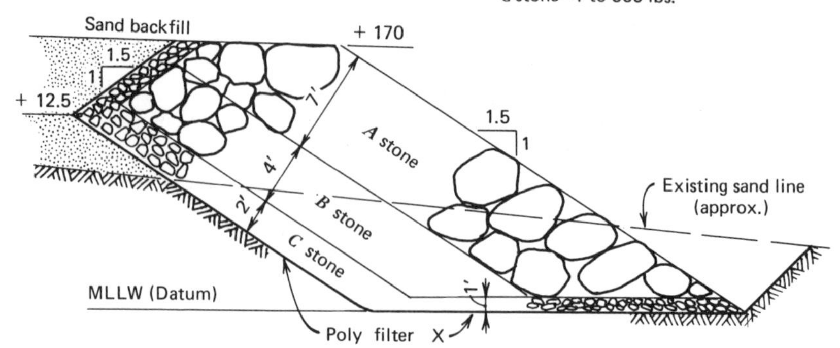

Note: Extend filter cloth at least 1 ft past edges and lap 2 ft at seams. Lay loosely, do not stretch.

FIGURE 6.2. Typical section—Coronado Shores Seawall (from Ref. 10).

Structures using construction fabrics and having a generally similar cross section have been constructed along the Atlantic seaboard since 1958. Naturally, weights, slopes, and elevations varied with local conditions.

Figure 6.4 shows the cross section of stone structure constructed by private interests in the Gulf of Mexico in 1970. The design was dictated by an absolute limit on the funds available to the owner. This limitation prohibited the use of armor stone as heavy as would normally be required or placing the toe to the proper depth. To overcome these restictions, a "sausage" was formed at the toe by wrapping 300-pound (140-kilogram) stones in fabric at -2 feet (-0.6 meter) with the expectation that, if under-

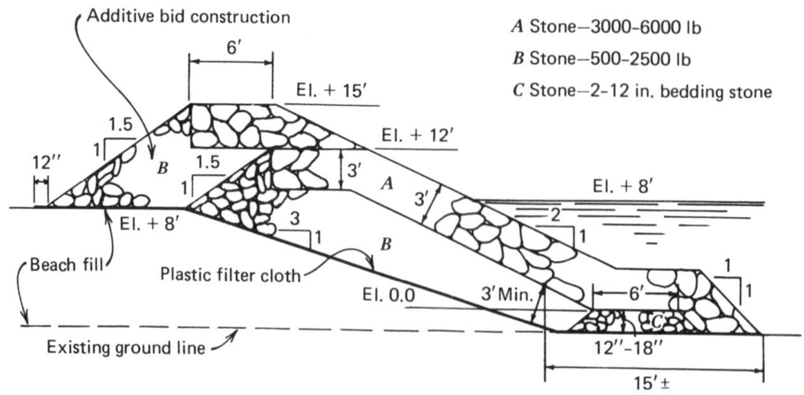

FIGURE 6.3. Typical section—Carolina Beach Seawall (from Ref. 10).

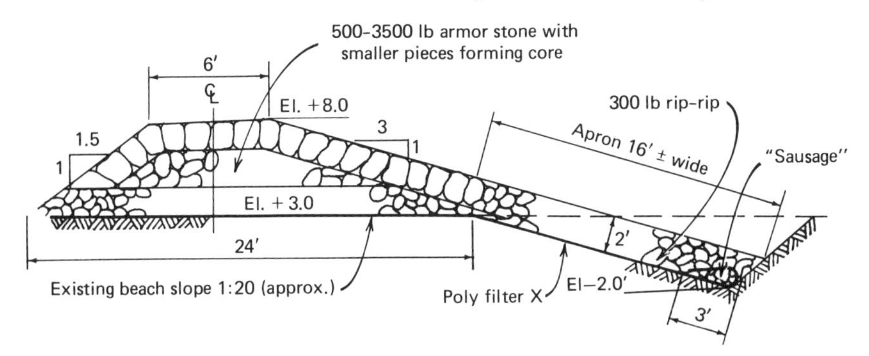

FIGURE 6.4. Typical section—East Timbalier Island Erosion Control (from Ref. 10).

mined, the sausage would sink as a unit until the toe was stabilized. Subsequent inspection over the past 3 years has proven this to be true.

After wrapping the sausage, the plastic filter was unrolled landward beneath the apron, and the base of the seawall and stone was placed upon it as shown. The apron provides a roughness factor that decreases the severity of the daily wave attack forces on the seawall, and the sausage maintains the integrity of the apron, which in turn prevents undermining that normally would occur with the base of the main structure at such a high elevation.

The structure has withstood hurricane tides and wave action from hurricanes. The owner has extended the seawall yearly. By fall, 1973 the entire Gulf side of the island had been protected with a structure of this design.

CASE 5. Bridge pier foundations in flowing water are particularly susceptible to water scour. Depths of scour of up to 36 feet have been reported,[11] and the topic is so important that the Highway Research Board had evaluated the state of the art in detail.[12] We feel that their findings on available design methods to predict scour depth were so erratic as to cast serious doubt on any of the available methods.

If bridge piers, or other structural obstructions, are necessarily founded in flowing water some positive action in preventing scour should be taken. One possibility is the use of fabrics. Hoedt and Mouw[13] present an interesting case history where a flexible ballasted fabric "skirt" was unfolded as a bridge pier was placed, (see Figure 6.5). The system developed by Enka b. v. of Holland comprises the unfolding of a flexible skirt, attached to the structure but nevertheless permitting the structure to be sunk into the soil to the required depth. The skirt is made of a strong sand-tight woven fabric.

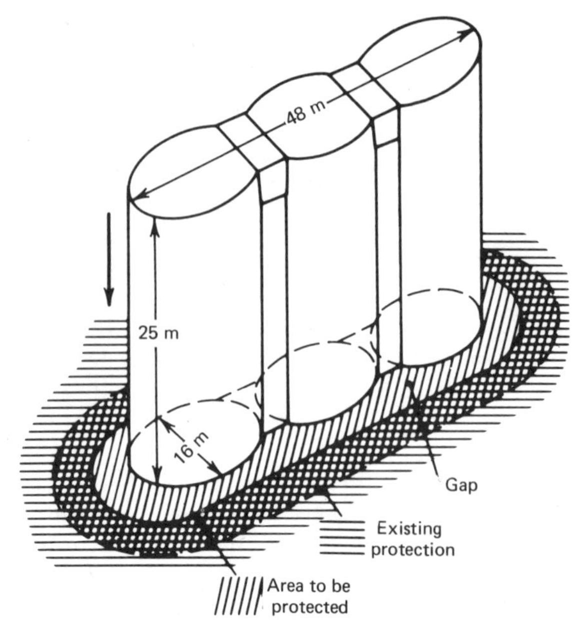

FIGURE 6.5. Pier foundation structure to be sunk into the subsoil; design dimensions are approximate (from Ref. 13).

Ballast, partly in form of rigid beams, is fixed to the fabric. The beams support the skirt in a vertical position against the side wall during transport and ensure that the fabric unfolds. The fabric width between the beams and the line of attachment to the wall is of the order of 5 meters. This part of the fabric can be folded up inside the wall. The functioning of the system is illustrated in Figure 6.6.

Depending on the required depth of a structure in its final position, there are two alternatives. The first is that the fabric width is sufficient to permit the fabric to remain attached to the wall (e.g., depths up to 10 meters). The second is that the attachment to the wall is such that the fabric is torn off after it has been lowered over a certain distance (e.g., 6 meters). In the latter case the rigid beams, when touching the wall, function as a barrier, preventing the fabric from being drawn down any further. The tension building up in the fabric will then trigger the tear-off mechanism.

The impression gained from the laboratory and field tests with this system is that the use of this prefabricated unfolding scour protection system need not be limited to the gated barrier piers. The system may well

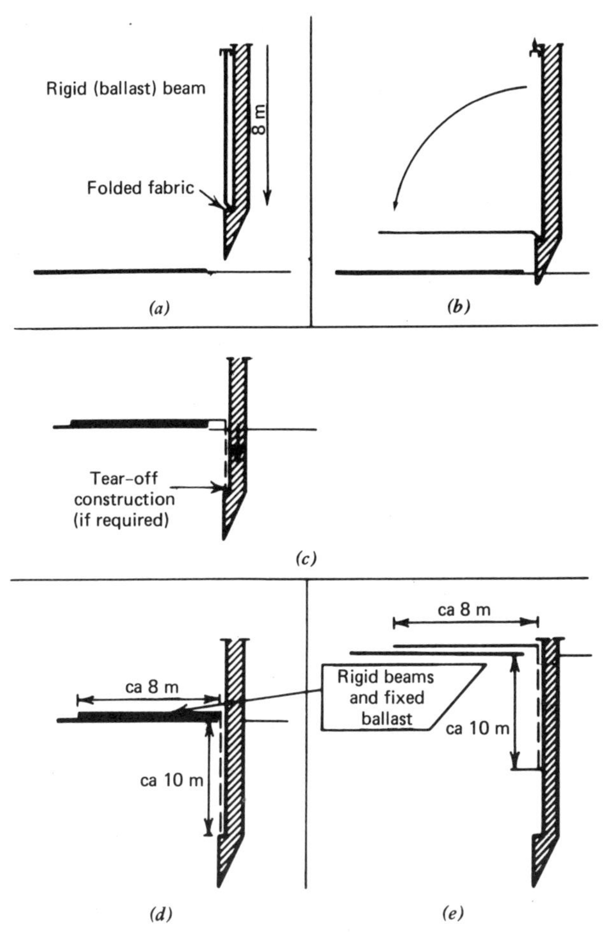

FIGURE 6.6. Functioning of the system. (*a*) Lowering of the structure; (*b*) unfolding the skirt; (*c*) sinking of the structure; (*d*) final position at small depth; (*e*) final position at great depth (from Ref. 13).

prove to be an economic way of protecting offshore constructions like gravity-structure drilling islands or bridge piers in rivers or tidal regions.

CASE 6. Similar in concept but quite different in actual construction detail is the ICI Linear Composites Scour Prevention System.[14] It has been successfully used on a number of offshore drilling platforms and loading terminal bases. The basic goals are to decrease the water velocity in the vicinity of the structural object and to cause a buildup of sand around the object. This is achieved by surrounding the structure with curtains of fibers in different geometric configurations. The lower ends of the fiber curtains are weighted down or anchored to the seabed, forcing the water to flow through the curtain, reducing its velocity and causing its suspended sediment to be released and deposited adjacent to the structural object. This type of system is discussed further in Section 6.4.

6.3 Erosion Control Mattresses

Background

As noted in the background portion of Section 6.2, the placement of revetment mats on slopes to be protected is one of the methods of erosion control. It is singled out here as a separate section because of the very active interest of fabric manufacturers and contractors in this area. Note should be made that the topic could also fit into Chapter 7, which deals with fabric use in forms, but is placed here instead due to its unique application as an erosion prevention system.

The mattresses generally consist of double layers of woven fabric forms placed on the slope to be protected and filled with concrete or grout. It will be seen that this type of forming system is a simple, fast, and economical technique for the placement of concrete for slope protection both above and below the water without the need for dewatering. Its performance characteristics and cost advantages make the process an adaptable and logical choice for stabilizing and protecting shorelines, levees, dikes, canals, holding basins, and similar erosion control projects.

The systems make use of the pressure injection of fluid fine-aggregate concrete into flexible fabric forms. Controlled bleeding of mixing water through the porous fabric produces all the desirable features of low water/cement ratio mortar—rapid stiffening, high strength, and exceptional durability.

For normal installations, the fabric forms, prefabricated to job specifications and dimensions, are simply spread over the terrain, which has received minimal grading. The fabric form is then pumped full of mortar. No granular filter blanket or filter cloth, as was described in Sections 6.1 or 6.2, is required.

This same concept can be used where slide problems are caused by eroding of the toe of the slopes, and where access is difficult for placement of rip-rap. Concrete-filled tubes have been installed and inflated to various lengths, heights, and thicknesses by pumping the concrete from an accessible area.[15] See Figure 6.7 for typical sections of mattresses and Figure 6.8 for typical sections of tubes.

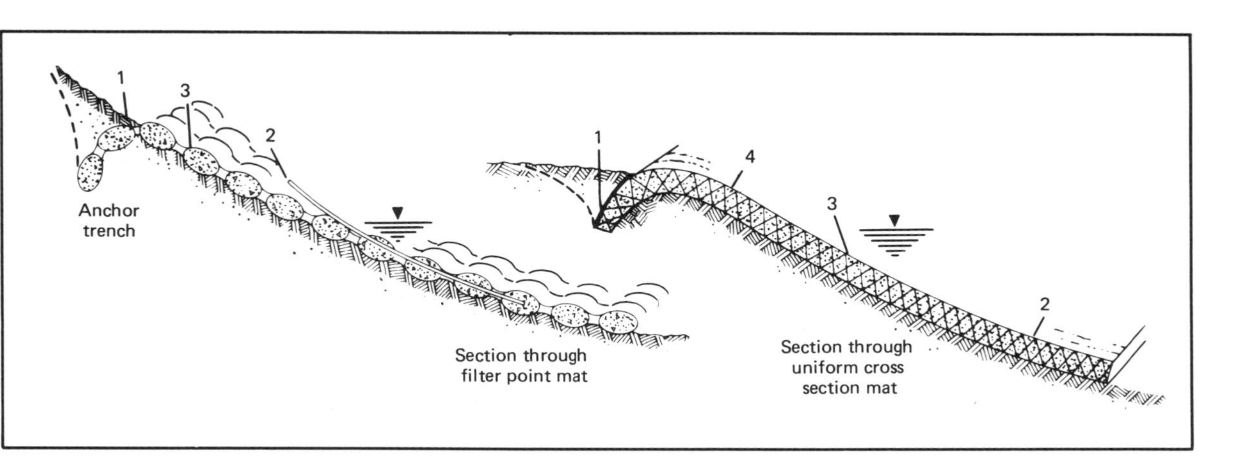

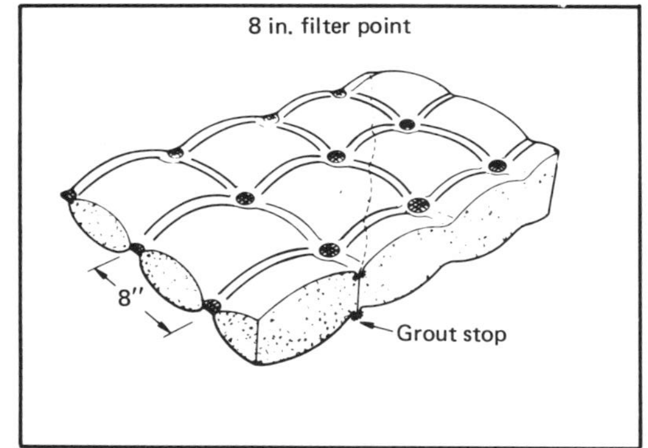

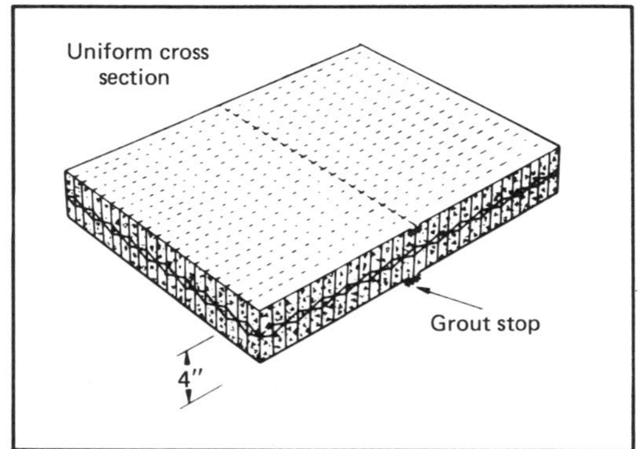

FIGURE 6.7. Typical cross sections and schematics of erosion control mattresses. Numbers in upper sketches refer to sequence of mortar injection (after Construction Techniques, Inc., Cleveland, Ohio).

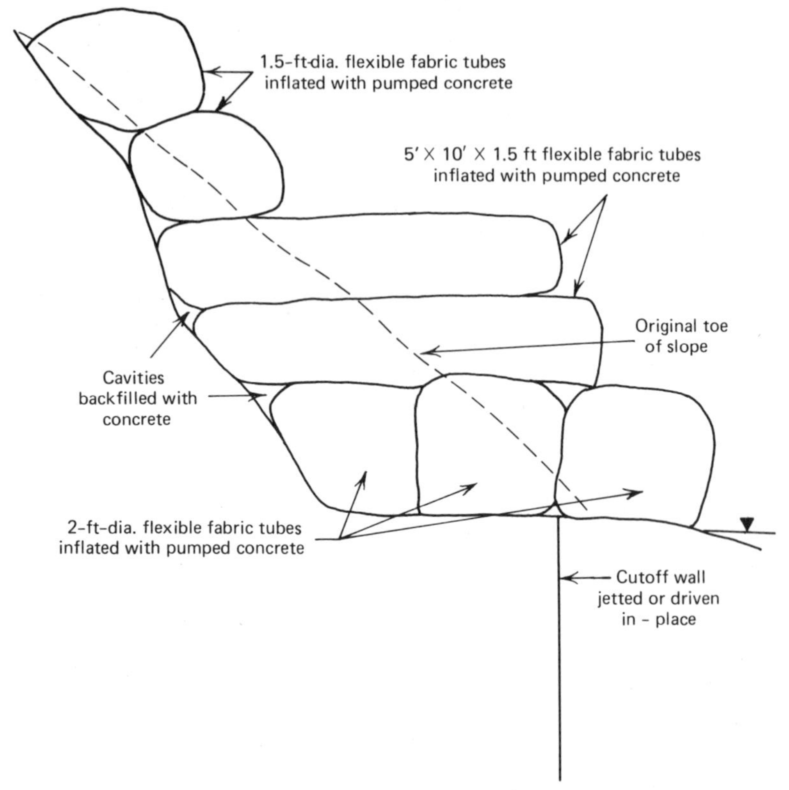

FIGURE 6.8. Typical cross section of erosion control tubes constructed at toe of slope (from Ref. 16).

Typical Installation Sequence

- Remove stumps, boulders, and brush from the site. Grade sufficiently to provide a slope that is stable in the absence of erosive forces. In general, an average slope steeper than 1:1 is not recommended. Cut an upper toe trench to prevent undercutting of the completed structure in event of heavy runoff.
- Place the shop-assembled fabric panels (usually from 2,000 to 3,000 square feet each in area) over the embankment, with seams straight and preferably perpendicular to the shoreline. The flexible, lightweight fabric is usually placed by hand, beginning upstream. Guide ropes to the opposite shore or small boats may also be used to assist in fabric placement.
- Sew or otherwise join the fabric panels, as delivered to the job, to-

gether to create a monolithic structure of any required length and width.

- Inject ready-mix mortar into the fabric envelope with a mortar pump; one with a capacity of from 10 to 12 cubic yards per hour is usually sufficient. The fabric in the anchor trench is pumped first to serve as a positioning function. Fill the underwater portion of the mat next. Then fill the remaining section of the fabric. The lateral spread of the fluid concrete can be controlled by shop-installed strips, which serve as grout stops. Production rates as high as 5,000 square feet per 8-hour shift are regularly achieved on large projects—with a team of just four laborers and one supervisor.
- Backfill the toe trench at the top.

Whereas many applications using fabrics are at the discretion and imagination of the user, many modern erosion control mattress systems are patented processes. They are as follows.

CASE 1. FABRIFORM. The Fabriform® system is a patented process of Construction Techniques Incorporated. Fabriform erosion control mats are produced in two basic designs, Filter Point and Uniform Cross Section. Examples of each style are shown in Figure 6.7. Fabric is normally woven in 60-inch mill widths and shop assembled to required panel size for field use. The fabric is woven of textured nylon fill in a multifilament nylon wrap. It is generally filled with the mix proportions of Table 6.1. Excess mixing water expelled through the permeable Fabriform fabric will reduce the volume of 27 cubic feet of wet mortar to about 23.5 cubic feet of hardened mortar (see Table 6.2.).

Pennsylvania Electric's Seward Station 215-kilowatt Fossil Plant is situated on the west bank of the Conemaugh River approximately 11 miles northwest of Johnstown, Pennsylvania.[16]

TABLE 6.1. Typical Mix Proportions of Concrete Fill

Material	Quantity Per Cu Yd
Cement	900–1,000 lbs
Sand (concrete or masonry)	2,200–2,000 lbs
Water	570–610 lbs
Water: cement ratio	0.63–0.61

Source: Ref. 16.

TABLE 6.2. Typical Yields of Concrete Fill

Material	Quantity Per Cu Yd
Original water	570–610 lbs
Lost water	218 lbs
Remaining water	352–392 lbs
Final water: cement ratio	0.39

Source: Ref. 16

The Johnstown area first obtained flood notoriety when the South Fork Dam broke on May 31, 1889, resulting in the death of over 3,000 people. This dam, located on the South Fork run, 14 miles east of Johnstown, feeds into the Little Conemaugh River, which joins Stoney Creek in Johnstown to form the Conemaugh River. The next major flood to hit the Johnstown area was caused by the severe 1936 winter. Much snow fell in January and February and very low temperatures prevented it from melting. Then in late February, an extremely warm period caused rapid melting of the snow and ice in the rivers and Johnstown and the Conemaugh River were once again flooded. Although property damage was high, loss of life was very small. This flood of 1936 was determined by hydrologists to be in excess of a one-in-two-hundred-year storm. The peak maximum discharge of the Conemaugh River at Seward Station was estimated to be 90,000 cubic feet per second. In the design of a new dike to surround Flyash Pond No. 2, it was necessary to be sure that the dikes were not only strong enough to take the force of any future flood waters but also higher than the two past major storms. The siting of Flyash Pond No. 2 was along the bank of the Conemaugh River about a mile southwest of the Seward Station, immediately downstream from the town of Robindale.

The owner was concerned that if another major flood occurred and if the Flyash Disposal area was not protected that the light flyash would be carried downstream and pollute the environment. The Seward Station produces over 100,000 tons of flyash per year while burning about 650,000 tons of coal. Accordingly, the height of the compacted earth dike surrounding the Disposal site was designed to have a top elevation of 1086.5 feet. Existing ground at the river side of the toe of the slope varied between 1060 and 1067.

The river side of the dike was designed to have a 2 on 1 slope and the face of the dike was designed to be protected by 8-inch filter point concrete-filled revetment mats. The top of the revetment mat was toed in to the dike at an elevation of 1081.5 feet, the height of the 1936 flood at this location.

Also, in order to prevent undercutting of the dike, the revetment was designed to have a two-foot toe into original ground at the bottom of the dike. The mat was approximately 1,000 feet long and covered just under 38,000 square feet of the surface area of the dike.

The mattress was installed in October and November of 1976. It consisted of two sheets of high-strength nylon, which were connected together on an 8-inch grid pattern. In addition to segmenting the mat, the connection points serve a dual purpose of acting as filter points to prevent buildup of hydrostatic pressure behind the mattress. The fabric form was brought to the site prefabricated to meet site conditions in approximately 10,000-square-foot segments. When laid on a slope, the fabric form was filled by pumping into it a specially designed concrete mixture. One of the features of the fabric mattress is that the nylon fabric is designed to allow water in the concrete mix to bleed through the fabric without allowing solids to escape. This lowering of the water/cement ratio produces up to 20 percent higher strength in the concrete in the mattress.

The concrete mixture utilized in this project consisted of 7 bags of cement, 100 pounds of flyash, 2,200 pounds of concrete sand, and adequate water to make a pumpable mix.

After the mattress was installed, the upper 5 feet of the dike were placed from an elevation of 1081.5 feet to 1086.5 feet.

On July 20, 1977 heavy rains inundated the Johnstown area of Pennsylvania, causing floods of a greater magnitude than the 1889 and 1936 floods. At the Seward Station site waters were 5 feet higher than in the 1936 storm and the peak discharge was over 97,000 cubic feet per second. The town of Robindale situated between Flyash Disposal Point No. 2 and the Seward Station was completely wiped out; fortunately, there was no loss of life. However, other residents of the area were not as fortunate, with over 75 known dead throughout the Johnstown area.

However, the fabric-formed concrete revetment on the face of the dike functioned as planned. Although some scour of the upper unprotected dike above the mat took place, the mat itself was in excellent condition except for some minor damage caused mainly by heavy debris in the flood water scraping the mattress.

See Figure 6.9 for typical examples of Contech's Fabriform mattresses.

CASE 2. The Great Lakes in the United States act similar to a system of reservoirs.[16] When the net supply to one of the lakes exceeds the outfall, its level rises. In 1974 Lake Michigan approached record heights because the precipitation in recent years over the Great Lake basin exceeded the basin average. The Corps of Engineers classifies the shoreline immediately north of Chicago as "high bluff-erodible." The rise in Lake Michigan and the

FIGURE 6.9. Examples of Fabriform® mattresses.

heavy waves generated by storms have caused considerable damage to properties sitting on top of the bluffs along this area. These storms have eroded the toes of the bluffs and caused slope failures, endangering many expensive private developments. The beaches between the toes of the bluffs and lake level average approximately 10 feet wide; therefore, they are relatively inexcessible to construction equipment, except for barge-mounted equipment off the lake.

In the fall of 1974, concrete-filled nylon tubes were used to restore the eroded toes of the slopes at three locations. Two projects were located in Highland Park, Illinois, and one project was in Lake Forest, Illinois. The bluffs at each of these locations were approximately 80 feet high and the toe had been eroded to a height of approximately 12 feet, causing slides which had not yet caused structural failure but, if allowed to continue, would have endangered the residences involved. All three design and construction procedures were basically the same. Utilizing hand labor, the trees that had fallen were cut away, a cutoff trench was excavated in the beach, and into this trench a tube was inflated, pumping the ready-mix concrete from the top of the slope. Three additional two-foot-diameter tubes were placed behind this cutoff tube and interconnected with steel cables; 4 per 30-foot section. On top of these tubes 2 layers of nylon bags measuring 5 feet long, 10 feet wide and, when filled, 1 foot thick were placed. Other tubes of smaller diameter were placed on top of these bags to conform to the actual height of the erosion, a maximum of 12 feet.

An 8-bag mix was utilized and approximately 1 cubic yard of concrete was placed per linear foot of toe rehabilitation. Despite severe winter storms, two of the three projects have suffered no damage. One of the projects had some minor undercutting after a severe winter storm and required some repair, but all three projects have served their function and eliminated slope failures and protected the properties on the top of the bluffs.

CASE 3. Karim[17] describes the use of such mats as scour protection around bridge piers, berms and on the bottom of the channel. In order to prevent undermining, both upstream and downstream ends of the fabric were trenched-in about 4 feet below the natural flow line elevation. The fabric mats were installed on a 2:1 side slope.

CASE 4. VSL HYDRO-LINING. Hydro-Lining Concrete mat is a product of VSL Corporation, headquartered in Los Gatos, California. Hydro-Lining consist of two plys of specially woven fabrics which act as a flexible framework to be filled with concrete. The thickness and configuration are controlled by internal spacer threads which are woven into the upper and lower fabric enclosures. The process was developed in 1965 and has since been successfully applied in many countries. The Hydro-Lining system and weaving of the fabric formwork has been granted patents throughout the world.

VSL Mats are available in three different styles:

Standard These mats are of constant thickness (10, 15, 20, or 30 centimeters) which is the length of the spacer threads between the two fabric layers. The thinner mats are usually filled with mortar while the thicker ones can be filled with concrete with aggregate size up to 15 millimeters. Mats of 5, 25 and 50 centimeters can also be ordered. Drain pipes, cut-off plates and joints can be incorporated into the fabric form before grout or concrete placement. This style is recommended for use on stable subsoils.

Slab The slab style has filter strips which are woven into the fabric form at regular intervals (usually 80 by 80 centimeters). When filling with mortar in the spaces enclosed by the filter strips, the mat deforms, thereby conforming to the ground surface. The usual slab thickness is 15 centimeters. This style is recommended for use on soils subject ot settlement and for coastal structures.

Mesh In this type of VSL mat only part of it is filled with mortar and thus represents a broken mesh appearance. The gaps in the mat permit the surface to be covered with soil and planted. The usual gap size is 22 by 22 centimenters. This style can be used in combination with the other two styles, which are sometimes placed below water, and the mesh type, placed above water on the exposed slope.

CASE 5. TERRAFIRMA. Terrafirma® is a registered trademark of Fibermarkers, Ltd., and is available from Foreshore Protection Property, Ltd., Sydney, Australia. The mattress system is made from Terrapakt fabric forms, which come in self-locking sections 57 inches wide and in any length. After placing and anchoring the forms, mortar is pumped in at low pressure, filling the mattress, which is sufficiently permeable to expell trapped water if placement is below the water surface. The fabric melds into the mortar, giving a lattice-work appearance. The vertical stitch points remain free of mortar, allowing water to pass through, thereby relieving hydrostatic pressure from the underlying *in situ* soil.

CASE 6. GOBIMAT. A system similar in concept to filter point mattresses (which are formed *in situ*) is the Gobimat®, a Dutch patented system available from Erosion Control Systems, Inc., Metairie, Louisiana. Gobimats consist of precast concrete Gobibricks® (8 by 8 by 4 inches with 35 percent open area) bonded to a filter fabric to form a mattress 4 feet wide by 12, 16, or 18 feet long. The mattresses are shipped to the job site, then lifted onto the prepared slope. (They weight 1,300, 1,750, and 2,025 pounds, respectively.) A one-foot overlap of fabric is left free on each side of the mat for joining to the adjacent mat or for anchorage.

The filter fabric is made from a combination of polyethelene and polyester monofilament yarns. The concrete blocks are glued to the fabric with a two-component polymethane by a special process. There is no bonding of concrete blocks to each other, which allows for mat flexibility in handling and placement on irregular surfaces and for filtration of the mat while in place (see Figure 6.10).

CASE 7. TERRAFIX. Terrafix® interlocking concrete block system is a product of Terrafix Erosion Control Products, Inc., of Rexdale, Canada. The system combines a layer of interlocking concrete blocks and a three-dimensional Terrafix filter mat placed directly on the prepared soil subgrade to be protected. Each block is positively interlocked to adjoining blocks by means of mortise and tenon connections. This allows for a fully articulated free-drainage system. A range of block types is available as are different filter mats for various subgrade conditions.

CASE 8. DURA-BAGS. So far this section on erosion prevention mattresses has progressed from uniform-cross-section mats, to filter point mats, to discrete blocks on a continuous fabric backing, to this system, which features discrete sections that are not physically jointed together. An example is Dura-Bags®, a patented product of Erosion Control, Inc., of West Palm Beach, Florida.

The bags are made from filter fabrics in a 60-inch width and 80, 120, and 156-inch lengths. When filled with mortar they weight 6,100, 9,100, and 12,000 pounds, respectively, and when filled with sand they weigh 4,300, 6,400, and 8,500 pounds respectively. Three different types of bags are currently available. Each is a 840-denier woven nylon fabric sewn together with a monocord polyester fiber. The difference is in the type of finish. One is unfinished, the second type has a black acrylic resin finish, and the third has a polyvinyl chloride finish. They can be placed for shore protections, as shown in Figure 6.11, and beneath the water to protect pipelines or other structures.

CASE 9. FABRICAST MOLDED BLOCKS. Fabricast is a registered trademark of Intrusion-Prepakt, Inc., of Cleveland, Ohio. The system consists of large bags of high-strength synthetic fibers used to cast large concrete blocks in place, either above or below water. Each bag has a self-closing inlet valve to accommodate insertion of a concrete pumping hose. More than one valve per bag is provided for bags longer than 20 feet. Typical bag widths are

FIGURE 6.10. Examples of Gobimats® in place. Photographs courtesy of Erosion Control Systems, Inc., Metairie, Louisiana.

FIGURE 6.10. *Continued.*

from 20 to 80 inches, which results in finished block thicknesses of approximately 15 to 55 inches. Uses are for ocean and lakeshore slope protection, groins, repair to breakwaters and dams, as underwater pipeline cradles, and as molded cribbing (see, e.g., Figure 6.12).

CASE 10. LONGARD TUBES. Rather than using discrete bags of finite length in forming the erosion control barriers, tubes of unlimited length can also be used. Efforts to form flexible sand-filled tubes were made as early as 1957, but were not very successful. Eventually, in 1967, a patent was granted to a Danish firm, Aldek A. S., in conjunction with the Danish Institute of Applied Hydraulics. Their system, as explained below, was further developed in 1970.[18,19] The franchise for the Longard Tube System in the United States east of the Mississippi River is held by the Edward E. Gillen Company of Milwaukee, Wisconsin.

An inner tube of 0.23-millimeter impermeable polyethylene is manufactured in 28, 40, and 69 inch diameters. The outer material is a woven, flexible, high-density permeable polyethylene, with uv stabilizing additives. The tubes are hydraulically filled with sand at the site by attaching a steel inlet drum to one end and a regulating outlet drum at the other end. A diaphragm pump mixes sand and water, this mix is then pumped to fill the tubes. A typical 40-inch tube, 330 feet long, can be filled in 3 to 4 hours. Fine sands to coarse gravels (up to 5 to 10 percent by volume) have been successfully pumped.

General configurations of the tubes are shown in Figure 6.13. Figure 6.14 shows photographs of the tube-filling operation under typical conditions.

FIGURE 6.11. Typical uses of Erosion Control's Dura-Bags.® Photographs courtesy of Erosion Controls, Inc., West Palm Beach, Florida.

FIGURE 6.11. *Continued.*

FIGURE 6.11. *Continued.*

FIGURE 6.12. Typical uses of Intrusion-Prepakt's Fabricast® molded blocks. Photographs courtesy of Intrusion Prepakt, Inc., Cleveland, Ohio.

FIGURE 6.12. *Continued.*

Examples of construction with longard tubes.

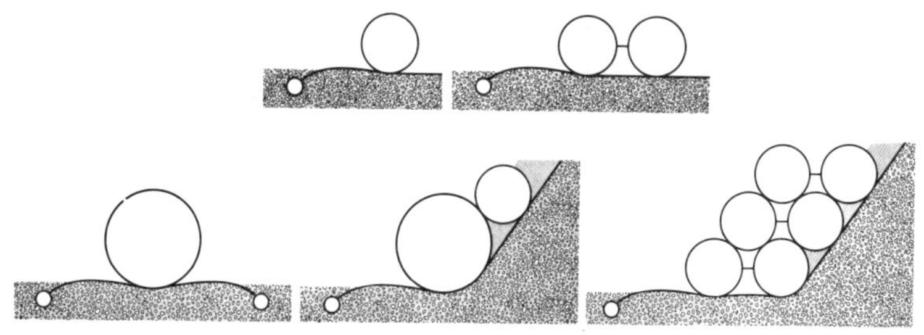

Filter fabric or a longard tube (10")/filter construction can be used under longard tubes if necessary.

FIGURE 6.13. Typical configurations of Longard tubes used for erosion control work.

6.4 Artificial (Fabric) Seaweed

Background

Inasmuch as vegetation slows the water runoff velocity on slopes to be protected on land, undersea vegetation can slow water currents to the point where all, or some, of the sediment being carried with it is deposited. By artifically creating such a situation under the sea at desirable locations a sandbank can be built up, thereby eliminating possible erosion from occurring. Since erosion and scour are wellknown hazards to underwater

FIGURE 6.14. Longard tubes under construction for erosion control purposes (from Aldek A. S., Denmark).

FIGURE 6.14. *Continued.*

construction, this type of application of fabrics seems to be quite appealing (see Figure 6.15).

Case Histories

CASE 1. Beginning in 1965 with an experimental planting off the coast of New Jersey, Avisun, a subsidiary of Sun Oil Co., has developed and tested an artificial seaweed called Olefern.[20] This material consists of fronds and strands made of 0.004 by 0.190 inch foamed polypropylene having a 0.6 density.

Testing of the material was done in the laboratory and in conjunction with NASA at Wallops Island, Virginia. The goals of the field study were to:

- Determine a satisfactory anchoring system.
- Evaluate the frond material.
- Assess the navagational hazard.
- Analyze the economics.

One year after installation it was observed that the Olefern spread in thickness and decreased in height. It resembled an underwater tumbleweed. The building of sand extended 6 to 8 feet away from the base of the frond. Various sea life (diatoms, gomphonema, barnicles, water blisters, and sea fern) and fish were also present in the area, which in previous years was noticeably unpopulated. Brashears and Dartnell[20] write favorably on all aspects of the project but caution against the adequacy of the anchoring system.

CASE 2. For approximately ten years, Nicolon artificial seaweed[21] has been used to slow down ocean currents, thereby decreasing erosion and fostering deposition of sand and the eventual stabilization of problem areas. Nicolon produces this material under an extruder gassing technique. The process involves stretching the polypropylene fiber foam strands which are ballasted to anchor the seaweed to the seabed. The seaweed remains upright in the water after it has sunk to the ocean floor because of its very low specific gravity. As a result, sand builds up in the seabed around the fibers and prevents further erosion.

Major tests of the material, reported at the International Conference of Coastal Engineering,[21] indicate that Nicolon polypropylene seaweed prevents erosion by tidal current once the artificial seaweed field is introduced. Other tests indicate favorable results surrounding the prevention of gullies in dune protection, in the prevention of bank slides, and in the protection of submerged pipelines.

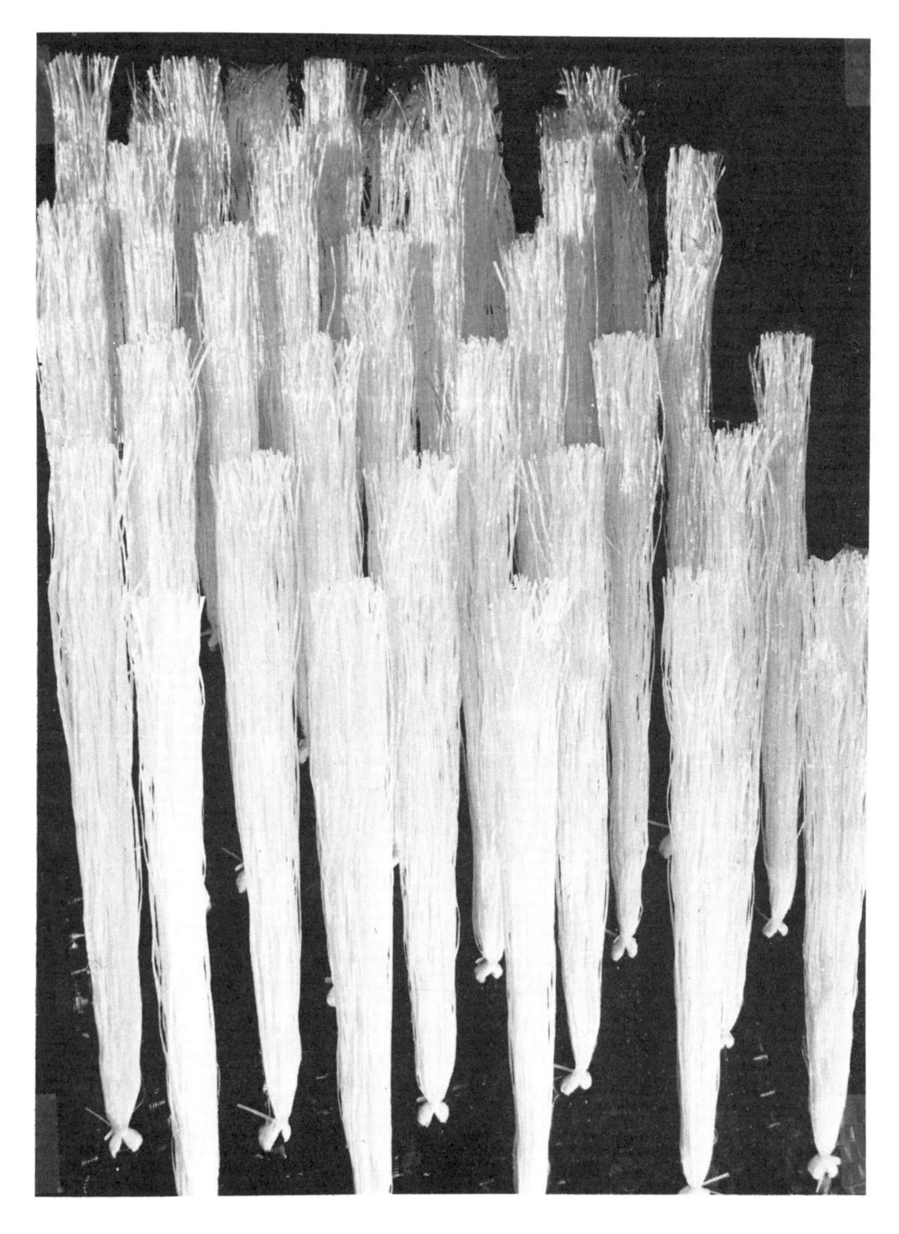

FIGURE 6.15. Polypropylene fronds (strands) secured to a filter fabric mat. Product illustrated is manufactured by Linear Composites, Ltd., a subsidiary of ICI, Ltd., England.

Regarding details of installation, the Nicolon artificial seaweed is applied generally in beds consisting of rows of light-weight foamed polypropylene tapes varying in length from $\frac{1}{2}$ to 2 meters with an anchor system that includes a hollow seam at one end of the tapes filled with a heavy material such as gravel. When placed on the ocean floor or seabed, the artificial seaweed promotes the buildup of sand.

Nicolon seaweed can be installed on beaches while the tide is out or it can be lowered from pontoons or supply boats to water depths of 35 meters or more. One of the most dramatic tests of the material was conducted off the Dutch Coast at the Leman gas field near two drilling rigs working for Shell and Esso in the North Sea. Thirty meters of the seabed underneath the pipeline linking the two drilling rigs were being eroded. A deep pit had formed where the pipeline rose from the sea floor to the drilling rig. To prevent a break in the pipeline, 1,300 sections of Nicolon seaweed were dropped overboard from a supply vessel. The curtain width was 2 meters, the seaweed length was 2.2 meters, and the water depth was 35 meters. After the material was laid it was inspected by divers 30 days later. The seaweed was reported buried in the sand, which had built up around it, and the pit had been largely filled.

Favorable results in other major installation examples are being produced with Nicolon artificial seaweed. These range from simple beach installations to a more complex steel beam system in which a beam is lowered from a pontoon previously filled with artifical seaweed. In this method, once the beam is lowered near the seabed, the pontoon is tipped over and the seaweed sinks and anchors to the ocean floor. The beam is raised, the pontoon is moved, and the operation is repeated.

CASE 3. A similar type of installation is reported[22] by Linear Composites, Ltd., an ICI subsidiary. The system consists of polypropylene strand bunches locked into a synthetic mat at half-meter intervals. The 1.5-meter-long strands float to form a curtain which reduces the water velocity, causing the suspended sand and silt to settle out and form a seabank. The bunches of polypropylene are locked between the warp and weft of the mat base so that they are protected from abrasion against the seabed. The base is weighted so as to keep it on the seabed. It is supplied in 4.4-inch-wide rolls.

The first full-scale trail began in 1968 when a prototype was laid alongside of a 46-meter-long pipeline in 9 meters of water. Within a few months a 1.5-meter sandbank had accumulated over the whole protected area. The pipeline was completely covered, although only a few meters away, where the system was not placed, it remained exposed. Occidental of Britian has recently installed a trial section on its North Sea Piper field oil line with good success of rapid buildup of accumulated sand.

CASE 4. In Delft, the Netherlands, the Royal Dutch Shell Plastics Laboratory has developed a polypropylene foam seaweed filament by mixing polypropylene with a gas, thus reducing its specific gravity to 0.2. The Shell seaweed is anchored by a woven curtain whose edges are filled with stones, gravel, or steel rods. The curtain can be adapted to the profile of the sea or river bed, reducing the risk of erosion below the anchorage. It is currently being tested in the Netherlands in an attempt to stop coastal and seabed erosion.

6.5 Fabric Silt Fencing

Background

Increasing interest in environmental concerns is dictating that special consideration be given to the water quality problems that are frequently caused by dredging, marine construction, or hydraulic landfill operations.

One of those problems is the turbidity that is often associated with such operations. The Water Quality Office of the Federal Environmental Protection Agency has issued *Water Quality Considerations for Construction and Dredging Operations*, which states that "Where silting and turbidity must be positively controlled because of the close proximity of the dredging operations to water intakes, resort areas, oyster beds, etc., a diaper or enclosure be provided for the cutter head and/or discharge pipe outlet to control the drift of suspended materials." In response, a growing number of federal, state, and local agencies are requiring the use of turbidity barriers for any job where the movement of suspended solids is likely to produce problems.

Depending upon the flow velocity of the water involved, various sized sediment will be transported. In most cases the particle size range will be 0.07 millimeters or finer, which is the silt and clay sized fraction of soils.

Fabric can be put to good advantage in "sieving" this material before it contaminates downstream waters.

While the discussion in this entire chapter on fabric use in erosion prevention has been addressed to erosion via water, there is also a significant problem arising from wind erosion. The types of soils susceptible to wind erosion are cohesionless sand and silts. In fact, silt erosion is so prevalent over major areas of the midwest that the area of loess soils is studied as a separate topic in geotechnical engineering. Loess soils are wind-blown silts resulting in a columnar structure, which is a reasonably stable *in situ* material but generally is not satisfactory when disturbed. In coastal areas where wind velocities are usually very high, sufficient forces

FIGURE 6.16. Typical uses of Erosion Control's Easy Fencin'.

FIGURE 6.16. *Continued.*

are mobilized to lift and transport these silt sized particles as well as larger sized sand particles. This soil migration can move massive amounts of material in relatively short periods of time unless arrested by some form of barrier. In the absence of natural barriers, artificial ones can be put to good use. Fabric fences are a possible alternative to solving these erosion problems.

Examples of fabric utilization for both water and air transported soils are given in the following case histories.

Case Histories

CASE 1. SILT-BAN. Silt-Ban is a product of Erosion Control, Inc. of West Palm Beach, Florida. The underwater fence is fabricated from a woven nylon fabric coated with polyvinyl chloride. It is installed in 100-foot lengths, of any depth, with flotation being provided by closed-cell foamed plastic logs contained in a heat-sealed sleeve of the barrier fabric. Ballast is used to keep the barrier vertical, with the main load line being furnished by a $\frac{1}{4}$-inch steel cable. Options depend on the water depth and are reflected in the size of the flotation logs. Two sizes are available: 3-inch-diameter logs for shallow draft and 6-inch-diameter logs for deep draft.

The system has been used in a variety of application areas as follows to:

- Form a settling basin to receive discharge from a disposal area sluice or weir.
- Surround a dredge to prevent movement of turbidity produced by the cutter or suction head.
- Form a settling basin to receive discharge or overflow from a hydraulic landfill operation.
- Surround or protect a recreation area or a valuable fish feeding or spawning area.
- Direct flows containing pollutants or suspended solids into areas where they will be carried away by currents.
- Function as a supplemental or emergency weirs in the event of dike leaks or failures.
- Form low-cost sequential settling ponds.
- Form a barrier around an open-water disposal area.
- Form an enclosure around marine construction or bank excavation work.

CASE 2. EASY-FENCIN'. Easy-Fencin' is a woven nylon fabric developed by Burlington Industries and marketed by Erosion Control, Inc., of West

Palm Beach, Florida. The nylon is dipped in acrylic resin for color and resistance to ultraviolet light degradation. See Figure 6.16 for typical applications.

CASE 3. MIRAFI 100X. As noted in Section 2.5.10, Mirafi 100X is a woven polypropylene fabric being used for silt fences and related applications. It is a licensed trademark and is manufactured by the Celanese Corporation.

6.6 References

1. S. P. Miller, "Bank Distress of Low Water Wiers on Big Creek, LA.," U. S. Army Engineers, Waterways Experiment Station, Vicksburg, Miss., Feb. 1978.
2. R. M. Koerner and T. A. Okrasinski, "Erosion Control of Granular Soils Using PVA," *J. Constru. Div. ASCE,* Vol. 104, No. C03, Sept. 1978, pp. 279-294.
3. "Soil Loss Prediction for the Northeastern States," Soil conservation Service, Upper Darby, Pa., 1962.
4. S. D. Rindge and D. A. Gaskin, "The Effectiveness of Two Erosion Control Fabrics in Retarding Soil Loss," Corps of Engrs. U. S. Army Cold Regions Res. and Engr. Lab., Hanover, NH, July 1977, 17 pgs.
5. ______ "New Stitches Sew Up Erosion Control," *The American City and Country,* May 1977, p. 73.
6. M. P. Keown, N. R. Oswalt, E. B. Perry, and E. A. Dardeau Jr., "Literature Survey and Preliminary Evaluation of Streambank Protection Methods," U. S. Army Engineer Waterways Experiment Station, Vicksburg, Miss., Tech. Report No. H-77-9, May 1977.
7. G. Baker, "Maine Department of Transportation Experience with Filter Fabrics," *Highway Focus,* Vol. 9, No. 1, May 1977, pp. 1-16.
8. J. Hoogendoorn, "A Case History of the Large Scale Application of Woven Synthetic Filter Fabrics on the Banks of the River Yssel," *C. R. Coll. Int. Sols Text.,* 1977, Vol. II, pp. 243-247.
9. R. J. Barrett, "Use of Plastic Filters in Coastal Structures," in Proceedings of the Tenth International Conference on Coastal Engineering, Tokyo, Japan, Sept. 1966, Chap. 62, pp. 1048-1067.
10. J. W. Dunham and R. J. Barrett, "Woven Plastic Cloth Filters for Stone Seawalls," *J. Waterways, Harbors Coastal Eng. Div. ASCE,* Feb. 1974, pp. 13-22.
11. K. Terzaghi, and R. B. Peck, *Soil Mechanics in Engineering Practice,* Wiley 1967, 2nd Edition.
12. "Scour at Bridge Waterways," National Cooperative Highway Research Project No. 5, Highway Research Board, Washington, D. C., 1970.
13. G. den Hoedt and K. A. G. Mouw, "The Application of High Strength Woven Fabrics in Hydraulic Engineering Constructions," 7th International Congress Koninklyke Vlaamse Ingenieursvereniging V. Z. W., May 1978, Vol. 2, pp. 11/1-11/9.

14. Product Literature, ICI Fibres, Linear Composites Scour Prevention Service, Hookstone Road, Harrogate, North Yorkshire, HG2, 80N, England.
15. Bruce A. Lamberton, "Revetment Construction by Fabriform Process," ASCE National Meeting on Environmental Engineering, Chattanooga, Tenn., May 1968, in *J. Constru. Div. Proc. ASCE,* July 1969.
16. J. P. Welsh, "Utilization of Synthetic Fabrics as Concrete Forms," Proceedings of Lehigh University International Symposium on New Horizons in Construction Materials, Nov. 1-3, 1976, Bethlehem, Pa.
17. M. Karim, "Concrete Fabric Mat," *Highway Focus,* Vol. 7, No. 1, June 1975, pp. 16-23.
18. R. Zirbel, "Sand-Filled Tubes Used in Beach Protection Plan," *World Dredging Marine Constr.,* Dec. 1975.
19. L. E. DeMent, "Two New Methods of Erosion Protection for Louisiana," *Shore and Beach,* Jan. 1977, pp. 31-38.
20. R. L. Brashears and J. S. Dartnell, "Development of the Artificial Seaweed Concept," *Shore and Beach,* Oct. 1967.
21. Product Literature, Nicolon Corporation, U. S. Textiles Sales Corp., 4229 Jeffrey Dr., Baton Rouge, La. 70816.
22. ———, "Polypropylene System Combats Pipe Line Scour," *Ocean Industry,* Dec. 1977, p. 103.

7

Fabric Use in Forms

Why stiff, rigid, inflexible formwork for concrete, grout, and soil fills? Probably because it is the traditional construction method. If we free our minds from such constraints, however, many previously difficult jobs are practical, and many heretofore impossible jobs can be done.

The use of a flexible and permeable form, that is, a fabric, has interesting possibilities. This chapter explores a number of them.

7.1 Fabrics as Concrete Forms

Background

It can easily be visualized that a highly permeable fabric in the shape of a bag or similar enclosure could act as a form for concrete or grout placement. Such a system could work equally as well above ground (where air is displaced from within the form) or in the water (where water is displaced from within the form). The resultant shape of the solidified mass would take the shape of the enclosure, which could be made in many different configurations. Equally important is that the empty form could be placed in difficult-to-reach locations and then filled after it is properly positioned. These various situations can best be visualized by way of actual projects, a number of which have been completed.

The major variables to be considered in this type of application of the use of construction fabrics are the following:

- Type of fabric to be used. This depends on the desired permeability of the fabric, the viscosity of the fill material, the ease of making joints and closures, the desired stiffness of the form before filling, the final shape of the solidified mass of concrete or grout, the cost, and the danger of patent infringement.

- Placement of the fabric form insofar as construction method, type of labor, inspection before filling, and so on, are concerned.
- Design of the concrete or grout filling the fabric form.
- Placement of concrete or grout in filling the form insofar as accessibility, time, cost, and so on.
- Manner of curing the concrete or grout after placement.
- Inspection of completed installation to check on the adequacy of the work and to see if further modifications are required.
- The setting up of a possible long-term monitoring system if desirable.

Case Histories

CASE 1. MISSION DAM IN BRITISH COLUMBIA, CANADA. As described in Section 3.1 on the use of fabrics in separation, the Mission Dam project[1] saw the development of two original and innovative uses of fabrics. This, the second use, has to do with using fabrics as forms to cut off seepage flow beneath the dam in the underlying pervious acquifiers.

As seen in Figure 3.2, the site is underlain by two pervious aquifiers. Under the major portion of the dam site, these are separated by a thick stratum of highly compressible clay. A deep grout cutoff through the lower aquifier extended to a maximum depth of about 520 feet, and a sheet pile cutoff through the upper aquifier controlled seepage through the foundation. The nature of the available construction material and the anticipated large differential settlement (up to 15 feet!) led to the design of a zoned type of embankment. The existing diversion dam had been in place for almost 10 years when it was decided in 1955 to raise the height of the dam by 140 feet. The diversion dam was designed to become an integral part of Mission Dam because it occupied the only site at which an adequate cutoff could be installed. The existing diversion dam had a length of 825 feet and a height of 60 feet above the riverbed. The diversion dam would serve as the upstream toe of the new Mission Dam. It was constructed of a rockfill shell with a thin vertical clay core, which extended as a clay cutoff to the top of the clay stratum. The presence of two acquifiers separated by the clay stratum required two separate cutoffs for the new dam. Site conditions required that both of these cutoffs be placed through the existing diversion dam. The lower cutoff consisted of a grout curtain extending from the bottom of the clay stratum to bedrock; the upper cutoff consisted of a sheetpile wall, which reinforced the original clay cutoff of the diversion dam.

If the clay core and cutoff had been intact and strong enough to survive an increase of the unbalanced water pressure on its upstream face equal to 140 feet, no further provisions against loss of water through the upper aquifer would have been required. However, the performance of the diversion dam during the first decade of its service showed that its clay core and cutoff were too weak to survive the increase of the elevation of the highest water level in the reservoir that would follow the construction of the Mission Dam. Therefore, it was decided to supplement the clay core and clay cutoff of the diversion dam with a new sheet pile cutoff. Like the core and cutoff of the diversion dam, the central portion of the replacement cutoff would settle many feet with reference to the rock-supported end sections, which would be subject to irregular warping in a downstream direction on account of great variation in the compressibility of its lateral support.

No thin cutoff could be expected to be subject to such deformation without serious injury unless it had the mechanical properties of a stretchable membrane, such as the wall made out of steel sheet piles with strong interlocks.

Most of the difficulties associated with the installation of the sheet pile cutoff grew out of the fact that the new cutoff occupied the site of an older rather deficient one and that the owners could not tolerate any interference with the operation of the existing reservoir. The clay core and clay cutoff of the diversion dam are located between cohesionless sediments and fill material containing boulders and cobbles. In order to avoid the necessity of driving the sheetpile through such unfavorable ground, it was decided to establish the new sheetpile cutoff within the clay core and cutoff of the diversion dam. The consistency of the clay in the core and cutoff varies between soft and very stiff but as a whole it is so stiff that sheet pile with a length of 100 feet could not be driven into it. In order to get sheet pile with that length into the ground, the contractor was compelled to install them in an open trench with a minimum depth of 60 feet. The trench consisted of a continuous row of mud-filled drill holes with a diameter of 20 inches.

The trench had to be kept open until the sheet piles were installed. Although the upstream wall of the trench was acted upon by the full water pressure of the adjacent reservoir, its stability was maintained exclusively by the side pressure of the drilling mud. For practical reasons, the density of the drilling mud was kept at or slightly below 1.25. Lower densities would have caused instability; higher densities would have prevented the subsequent replacement of drilling mud by grout.

As soon as the section of the 20-inch trench was completed, the sheet piles were installed in the central portion of the cutoff and were driven to an imbedment of 20 to 40 feet into the clay stratum. Then a 7- to 27-

foot panel of mud-filled trench was cleaned of sediment mud by churning the chisel jets with diameters ranging from 2 to 12 inches. Loose fragments such as chunks of hard clay or rock chips were removed from the bottom of the trench by special air-lift pumps. After cleaning, the drilling mud and the panels were replaced by grout with density of about 1.8 consisting of a mixture of clay, cement, sand, and barytes.

It was at this point necessary to separate the 7- to 27-foot-long panels between the bentonite slurry and the cement-bentonite grout. Terzaghi designed nylon bags known as sausages, which would conform to the irregular spaces between the clay core and the inner steel sheet piling. However, to assure that this technique would be adequate, a multicurve concrete surface about 20 feet high was built against the curved wall. Steel sheet piles were installed vertically several inches from the curved wall and a nylon-reinforced plastic canvas bag with a length of about 20 feet and a diameter of about 18 inches was wrapped around a 1-inch-diameter grout pipe. The pipe and bag were introduced between the multicurved concrete surface and the sheet piling. The bag was filled with cement-bentonite grout.

After the test had proved satisfactory, the 65-foot-long nylon sausages were lowered into place, filled with specially-formulated cement bentonite, and successfully acted as a vertical barrier between the adjacent bentonite slurry and the cement-bentonite grout. The project was highly successful and proved to be the first use of construction fabrics as grout or concrete forms.

CASE 2. UNDERPINNING BRIDGE PIERS. In the fall of 1968 the Pennsylvania Department of Transportation (PennDOT), while performing a bridge inspection, encountered a void beneath Pier 16 of L.R. 25, Northumberland County.[2] This 28-span through girder bridge crosses the Susquehanna River between Sunbury and the Shamokin Dam in Pennsylvania. A recreational dam was planned immediately downstream from this bridge. The dam would raise the water level to a greater depth and change the flow characteristics of the river. Because of these facts it was elected to repair the underscour problem immediately.

The Susquehanna River at this point is approximately 2,500 feet across, and the bridge piers are on rock. These conditions eliminated access roads, cofferdams, or sheet piling alternatives from consideration. Therefore, PennDOT elected to try the technique of placing concrete underwater by the utilization of a permeable fabric form. In effect, a woven nylon tube was designed to fit into the void caused by scour between the top of the rock and the bottom of the pier foundation. This tube was pressure filled with fine-aggregate concrete and extended into the void as much as possible and partly outside of the pier. Prior to inflation, pipes were placed into the void beneath the pier and, after the tube of nylon was inflated with the pumped concrete, the same concrete mixture was then injected into the void space behind the tube (see Figure 7.1). Sufficient pipes were placed so that water could be vented out from the void, thus assuring that complete filling beneath the pier was accomplished.

In June of 1972 Hurricane Agnes rampaged through the mid-Atlantic States. One of the hardest hit areas was the Susquehanna Valley of central Pennsylvania. A considerable number of bridges were destroyed, and subsequent investigations by PennDOT indicated that many other bridges had experienced scour problems beneath their foundations. In the course of the bridge-damage inspection, the Northumberland County bridge was reinvestigated, and it was found that the tube of concrete was in

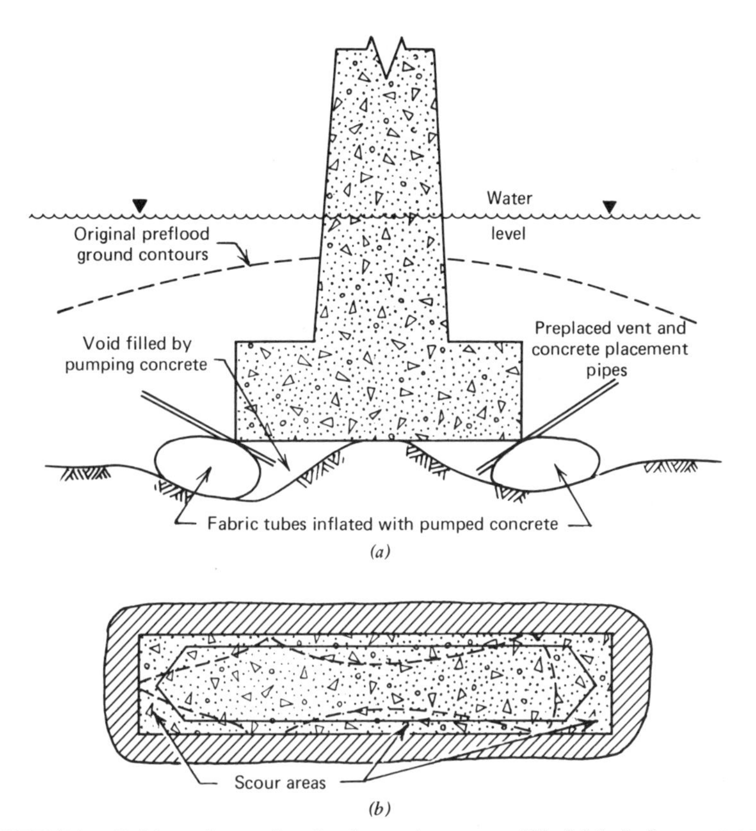

FIGURE 7.1. Bridge pier underpinning using grout-filled fabric forms. (*a*) End elevation, (*b*) plan (after Welsh, Ref. 2).

place and no further scour was experienced at this pier. However, eight other piers of this structure had experienced damage from the velocity and volume of the water that had poured down this valley, and these piers and other piers and abutments were repaired by the fabric form techniques. In 1975 Hurricane Eloise hit the same area of Pennsylvania and, once again, flows exceeded the one-in-a-hundred-year prediction. Additional diving inspections showed that structures previously repaired by the fabric form technique had experienced no further major scour problems.[2]

Caution is required in the use of this technique in that the underlying foundation material must be firm or relatively unyielding since the deformation ability of the underpinning is governed by the relatively inflexible characteristics of the unreinforced concrete.

CASE 3. UNDERPINNING CAISSONS. Hoedt and Mouw[3] have recently used four different types of fabrics for concrete forms beneath caissons in the construction of a gated storm-surge barrier in the Eastern Scheldt of the Netherlands. The barrier consists of 80 piers, between which movable sluice gates are suspended to control tidal variations of 1,100 million cubic meters of water at mean tide. In positioning the prefabricated caissons of a stoney sill, the following aspects were of concern:

- Support reactions.
- Correct vertical positioning.
- Seepage control.

A schematic of the problem and the solution using a fabric form between the bottom of the caisson and the stone sill is shown in Figure 7.2. Fabric specifications called for

- breaking strength of 250 kilonewtons per meter of width in all directions.
- 25 percent elongation or more to accommodate the irregular base.
- high cement retention.
- reasonable water permeability.
- large width (approximately 20 meters) to avoid seams within compartments.
- high tear resistance to avoid puncture by sharp stones.
- rotproof fibers because of long anticipated construction time.

Enkalon was the fabric eventually chosen (on the basis of extensive laboratory tests[3]) and was tested in a number of large-scale experiments

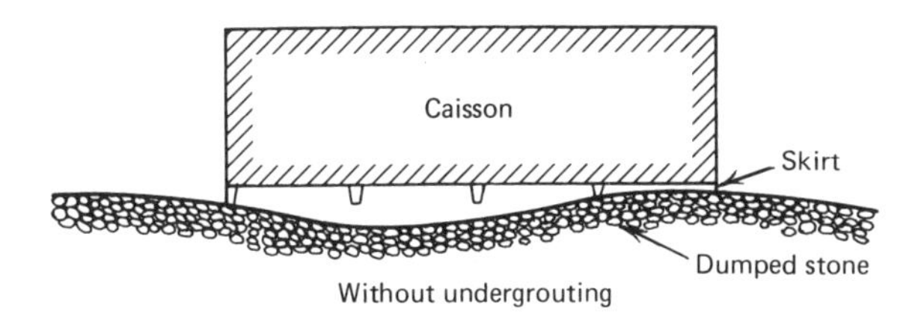

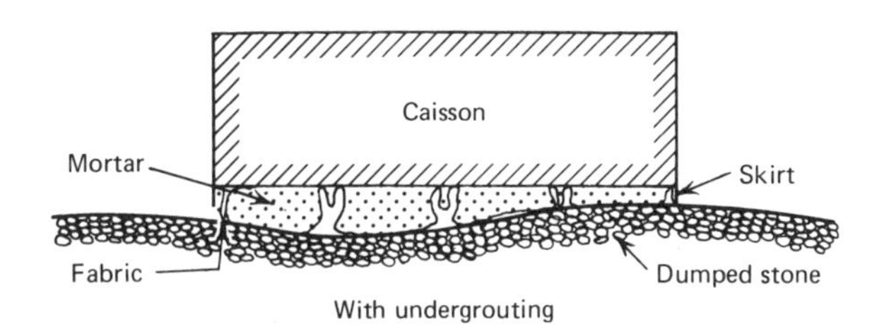

FIGURE 7.2. Test foundation at Kats, The Netherlands, showing the undergrouting and embedded stones (from Ref. 3).

with caissons measuring 7 by 6 meters, (see Figure 7.2). The project is currently under construction.

CASE 4. CUSHION FOR UNDERWATER TUNNEL. The firm of Christiani and Nielsen of Copenhagen has designed and supervised construction of a grout-filled fabric cushion for beneath the Tingstad Tunnel in Gothenburg, Sweden.[4] Originally designed to rest on a prepared bottom in a dredge trench, the steel-encased concrete structure instead had to be supported by piles. In a novel procedure, the contractor put a cushion of grout contained in large fabric bags between the tunnel and the piles.

The original intent was to float the tunnel on the deep clay of the river bottom by dredging it out enough to place a 10-foot layer of sand. But silt on top of the clay at the design elevation was found to be too unstable. Thus over 1,000 wooden piles had to be driven more than 70 feet deep to support the tunnel. The builders capped a cluster of six piles at each corner of each tunnel section with tremie concrete.

To transfer the tunnel loads to the piles the contractor, Skanska Cementgjuriet of Stockholm, devised and patented a grout cushion consisting of huge nylon sacks pumped full of grout to form-fit between the tunnel bottom and the top of the piles. The sacks are roughly 100 feet long and 13 feet wide. They inflate to about 7 inches thick between the piles and tunnel bottom; they inflate to twice that thickness if unconfined (see Figure 7.3).

Water pumped into 13-foot-diameter temporary steel cylinders sank the tunnel sections to the bottom after they were towed into place. Each section rested on the corner pile clusters, above the previously placed empty grout bags. The contractor then filled the bags with grout.

CASE 5. FABRIC TUBES FOR LNG OFFSHORE FACILITY. The $300 million LNG Facility located at Cove Point, Maryland, on the Chesapeake Bay, featured many unique construction problems and solutions.[5] Environmental considerations dictated the use of a sunken tube (i.e., a tunnel) to transmit the liquid natural gas from the one-mile offshore shaft to the onshore storage facility.

The last tunnel section installed connects to the offshore shaft underneath the operating platform. The original design called for the placement of large armor stones as the final protection cover for the tunnel. However, with the operating platform in place and with limited clearance beneath the operating platform this would mean that floating equipment would have to place these large stones from outside the perimeter of the operating platform. The contractor was concerned with the

FIGURE 7.3. Grout-filled fabric mats beneath Swedish tunnel (from Ref. 4).

possibility of damaging the 54-inch precast cylinder piles that support the operating platform during the placement of these large stones, which would literally have to be thrown underneath the structure.

For a safe alternate to the armor stone placement operation, two layers of large-diameter tubes inflated with concrete were designed and installed. Past experience has shown that fabric forms, when pumped with concrete, form a shape where the height is roughly one-half the width. Therefore, a design was developed where tubes of concrete up to 70 feet long with a height of 3 feet and a width of 6 feet were alternated with and adjacent to tubes 2 feet high and 4 feet wide. A second row was alternated with this design to interlock the structure and form a relatively smooth surface.

In the summer of 1976 the contractor mobilized for placement of these tubes and set up on the small boat dock immediately adjacent to the operating platform an 8-cubic-yard concrete agitator. The concrete mix used was 8 bags of cement, 2,200 pounds of concrete-sand, approximately 60 gallons of water, and 6 percent entrained air. As the on-site batch plant was previously dismantled, the concrete supplier loaded the dry portions at his plant 50 miles away and added water at the job site. The ready-mix trucks discharged this concrete into two 4-cubic-yard concrete buckets mounted on a barge, and a tub transferred this barge 1 mile to the offshore facility, where the concrete was pumped into the agitator. Divers took the predesigned nylon fabric tubes 30 feet to the bottom of the bay and positioned them with cables at both ends. The nylon tubes had self-closing injection points, and the concrete was then pumped through 2-inch hoses into the tubes. The tubes were built up on alternate sides of the tunnel to allow the concrete to set prior to placement of adjacent tubes. Over 975 cubic yards were injected into the 50 tubes by this method.

7.2 Fabric Forms for Pile Jacketing

Background

All piles in a marine environment suffer deterioration at varying rates. The deterioration is caused by normal marine exposure, wet/dry cycles, freeze/thaw cycles, and chemical, industrial, and sanitary wastes. Moreover, each type of pile has its own particular problems.

WOOD PILES. Wood piles in a marine environment can be subject to attack by bores. There are three basic types: teredo and bankia (the molluscs), and the arthropod limnoria. Although the limnoria is a surface eroder and the molluscs are internal borers, it is extremely difficult to detect damage to woodpiles by visual inspection. BC Research, Vancouver, British Columbia, has developed a nondestructive sonic testing method of determining the extent of damage by marine borers. (It is somewhat ironic that as we clean up our harbors and waterways and remove the pollutants, marine borers move in and cause extensive damage to these previously pollution-protected structures.)

CONCRETE PILES. Deterioration in concrete piles can be limited to the water and splash fluctuation zone if caused by wet/dry, freeze-thaw cycles. However, some concrete piles (both precast and cast-in-place) have been deteriorated to below the mudline from poor original placement techniques. This comes about from permeable concrete, which allows corrosion of the reinforcing steel to occur and subsequent spalling of the concrete. It is generally caused by a sulphite reaction of the concrete.

STEEL PILES. All types of steel piles are subjected to corrosion (average corrosion rates are 0.005 inch per year under normal conditions but can increase drastically under certain stress conditions) and damage by electrolysis. The latter can be prevented by cathodic protection methods, but these are costly to employ.

General Methods of Repair

The methods used for rehabilitation of piles are varied and constantly increasing in number. The oldest technique is to use metal forms, such as corrugated steel in half sections joined together by angles, attached to each of the half sections. Dockworkers and divers place and join the sections, which are used to contain cement grout, which bonds to the deteriorated pile section. This highly labor-intensive operation has led to development of other more economical techniques. Recently plastic forms have been proposed as not only an economical form but as a high-strength method of preventing further deterioration to the piling system. The annular space between the form and pile can be backfilled with either concrete grout or specially formulated epoxy. Another recent pile jacketing technique utilizes bituminized fiber forms. As can be envisioned, all of these systems have a problem with bottom closure when they are acting as a form and the pile is not to be jacketed down to the mudline.

Built-up layers of epoxy or bituminuous coatings have also been utilized to protect pilings from deterioration; however, most of these techniques involve hand placement by divers, and a thorough coating of the piles is difficult and expensive to obtain.

In the 1960s a technique was developed that utilizes construction fabrics as a concrete-forming system. Basically this concept uses as a concrete form a jacket of woven nylon fabric, each end of which is connected by a heavy industrial zipper sewn onto the fabric. The ends of the fabric above and below the deterioriated pile zone are banded to the pile. These fabric forms have proven economic advantages over other concrete forming systems because of their light weight, ease of installation, relatively low fabric cost, and ease of connection onto the piles at any location above the mudline. Basically the fabric is so designed that when concrete is injected into it the excess water bleeds through the weave of the fabric without allowing the cementatious portion to escape. This

FIGURE 7.4. Pile rehabilitation process using the Fabriform® jacket system, showing wooden piles almost completely decayed with reinforcement and jacket being installed, zipper being closed by diver, and completed section of dock (photos by B. A. Lamberton, Intrusion Prepakt, Inc., Cleveland, Ohio).

lowering of the water/cement ratio produces an extremely dense surface of concrete to resist further deterioration of the pile. See Figure 7.4 for typical installation procedures.

In 1969 Dillingham Corporation[6] demonstrated the advantages of this technique. Using 10 specimens of Grade A 14-pound creosoted Douglas fir piling with deterioration varying in the piles between 0 and 100 percent, it was found that a deteriorated pile jacketed by fabric-filled concrete would carry a higher load than the undeteriorated control piles.

The increase in the load-carrying capacity varied from 10 to 30 percent. It was also found that the weak link in the system was the timber pile itself, while the concrete jacketed section showed no evidence of excessive strain and no visible cracks developed. Furthermore, it was obvious from the tests that the load is transferred from the top of the pile through the concrete jacket and back to the bottom of the pile. The load transfer is accomplished by a mechanical bond between the concrete and the wood. The deflection of a jacketed timber pile under vertical load is not excessive in comparison to an undamaged timber pile. Welded wire cages were used in some of the jacketed concrete piles; however, this steel had little effect on the vertical load-carrying capacity of the jacketed piles.

Specific Techniques for Installation of Fabric Forms

Since this type of fabric jacket system relies on load transfer by mechanical bond, proper surface preparation of the pile is vital to its success.[7–9] Wood piles should be scraped of all marine growth and rotted timber removed. Loose rust and scale should be removed from steel piling by either wire brushing or underwater high-pressure water blasting. Sand blasting to bright metal is not required. The removal of loose and deteriorated concrete from concrete piles can be accomplished by underwater air hammers or underwater high-pressure water blasting. Exposed reinforcing steel should be cleaned and, if deteriorated, splicing of new steel accomplished. If the pile is to be jacketed slightly below the mudline, this can be accomplished by jetting out a deeper hole, and after the pile jacket is zippered together above the water, the diver works the jacket to the bottom and then bands it under the mudline with a banding tool. If jacketing is to be deeper into the mudline, then jetting out of the area and attaching a metal bottom form is recommended. A bottom seal of grout can be injected to prevent future leaking of the pumped concrete grout. Where deterioration does not extend the full length of the pile, the jacket can be installed at any location by a double banding of the bottom to the pile and then injection of a bottom seal. The top of the pile jacket can either be opened or closed. If it is opened, a ring is installed in a seam at the top of the fabric and this ring is supported from anchors in the superstructure or in the pile itself. When using this technique, a sloping seal of epoxy or drypack should be placed to prevent water from laying on top of the pile. The other and more straightforward method of supporting the pile jacket is to band the upper portion of the form at a higher elevation than where the deterioration has taken place.

Unless the piles are being designed for additional bending moment, a heavy reinforced cage of steel is not required in the annular space between the deteriorated pile and the outer portion of the jacket. However, in areas where freezing and thawing will be experienced, temperature steel consisting of welded wire mesh can be installed around the deteriorated pile prior to installation of the fabric form. The diameter of the pile jacket should be a minimum of three inches larger than the outer surface of the original pile.

The majority of the fabric pile jacketing material to date has been woven nylon. With this type of fabric an expansion of 6 to 8 percent of the area can be expected. The concrete is injected into the space between the pile and the fabric form by two plastic grout pipes inserted prior to the placement of the fabric. These pipes are placed through an open portion in the fabric. This is either through inserts or through slits in the fabric. Normally, two pipes are simultaneously injected with concrete. A normal mix design will utilize 8 to 11 bags of Type I or II Portland cement, 2,200 to 2,500 pounds of concrete sand, 50 to 60 gallons of water, and a pumping aid designed to impart 4 to 6 percent air entrainment and eliminate bleeding of the concrete. Caution has to be exercised with this technique when a strong current is prevalent, as the fabric form inflated with fluid concrete can be pushed off center from the pile. Also, the zipper adds rigidity to the side of the pile that it is attached to, and particularly on larger diameters the pile will tend to expand further on the side opposite the zipper closure. With this system of pile jacketing, as with any other technique currently available, difficulty has been experienced where permanent cross bracing is in the zone to be jacketed. On one recorded job, where a pile group was on fairly close centers, the form was designed similar to a three-legged diaper and the piles were jacketed simultaneously.

Costs

In most instances this technique of jacketing piles with a fabric form, whether utilizing a patented system or a self-designed technique, will prove to be the most economical method of pile jacketing because of its convenience and high productivity. The actual cost on each job will depend upon the number of piles to be jacketed, the diameter of the piles, the shape of the piles, the tidal fluctuation, labor conditions, the velocity and contamination of the water, and exposure of the pier to wind or waves.

Case Histories

CASE 1. CONCRETE PILES ON LAND. In 1968 the New Jersey Turnpike Authority awarded a contract to repair ninety 13-inch concrete octagon piles on Bridge 0.42. These piles were jacketed with 24-inch fabric jackets for an average length of 10 feet. Costs were approximately 15 percent less than were other methods. A site visit in 1978 showed no further deterioration of the piles or of the grout jacketed areas.

CASE 2. STEEL PILES ON LAND. In 1969 an oil refinery in New Jersey had fifty-two 12-inch BP53 steel piles jacketed with 24-inch fabric jackets for an average length of 6 feet. These pile jackets have performed excellently through the intervening years and additional piles will be similarly jacketed beginning in 1979.

CASE 3. CONCRETE PILES IN WATER. A *T*-shaped pier extending into the Arthur Kill River in New Jersey is used to unload containerized freight from large ships. The 18-inch octagon concrete piles supporting the pier were rapidly deteriorating. Some were so weakened that the concrete could be removed easily with a high-pressure water jet used for standard cleaning. In 1977, 75 piles were rehabilitated using 24-inch fabric jackets of a 10-foot average length. Approximately one-half of the jacket length was beneath the water surface. An additional seventy-five piles were similarly rehabilitated in 1978 and the remaining piles of the piers are scheduled for completion in 1979. Pile jacketing costs are approximately 15 to 20 percent less than competing pile rehabilitation systems.

CASE 4. PILE PROTECTION AGAINST CHEMICAL ATTACK. The concept of pile jacketing can be extended from using fabrics as grout forms to using fabrics as coating materials on the piles themselves. The fabrics are usually treated, making them impermeable to resist salt water corrosion and chemical attack. Frankipile, Ltd.[10] has used a PVC-coated nylon to protect concrete piles from sulphates and chlorides in the ground below the water table.

CASE 5. REDUCTION OF DOWNDRAG (NEGATIVE SKIN FRICTION). Downdrag (or negative skin friction) occurs on predominately end-bearing piles extending through compressible soils.[11] As the compressible soils settle they transmit forces to the piles, adding to the design loads. Single pile forces of greater than 100 tons have been measured, these forces causing pile failures in many cases. A bitumen slip layer has been used to coat the piles, but the technique is quite labor intensive and could be replaced by using a low-friction fabric. The fabric being placed around the pile before installation must be strong enough to resist tearing, but once in place should provide for an adequate slip surface. Research in this potential application of fabrics to pile systems seems justifiable.

7.3 Columns For Mine and Cavern Stability

Background

In many parts of the world abandoned underground mines and limestone cavities have caused major structural subsidence problems. Many technical articles have been written on alleviating this problem, and Gray, et. al. report, *State of the Art of Subsidence Control,*[12] discusses some of the techniques available. These include hydraulic flushing, grouting, grout columns, and so on. Quite frequently, in abandoned mines and underground limestone cavities sufficient cap rock or overburden is available to support the structure if the roof of the mine can be supported to eliminate subsidence and the relatively expensive technique of filling the cavity or mine is not required.

Construction Fabric Application

In those instances where columns for roof support are advisable, construction fabrics can be used as a form without the necessity of entering the mine or cavity. The technique consists of drilling 4 or 5-inch-diameter holes to intercept the roof of the mine and then carrying these holes to the floor of the mine, penetrating the floor approximately 1 to 2 feet. A tube of fabric, normally woven nylon, of a predetermined diameter is wrapped around a grout pipe and snaked down the drill hole into the key at the floor of the opening. Then fine-aggregate concrete is injected under controlled pressure as the grout pipe is withdrawn. The tube of fabric has to be supported at the surface or through rings at the top of the fabric with a cable system to the ground. Each application requires a determination of how much pressure the fabric can withstand, and it may be necessary to pump the tube in multiple lifts. Reinforcing steel can be placed either in the void area only or for the full length of the column. The critical point in this application is to get maximum support of the column of concrete at the roof of the mine. Where the cavity or mine is in the dry, it is feasible to snake down an adjacent hole and observe the final product with a television camera. In areas where the opening is fully or partially filled with culm or other compressible or objectionable materials, it is possible to jet out an opening in this material by having the grout pipe extend through the bottom of the tube and, while jetting through this pipe, maintain an adequate head of bentonite or grout in the fabric form. This technique has the advantage over other methods of forming grout columns in that a positive form can be relatively economically installed, as a considerable quantity of fine-aggregate concrete is not expanded in a large wasteful base to build up the angle of repose of the grout to the roof of the mine. Using this same concept, a bulkhead can be created in an underground mine by drilling holes on a predetermined line and pumping alternate columns initially and then placing the fabric in the secondary or intermediate locations and blowing them up with concrete to interlock between the original placed columns. Two walls can be created to form a bulkhead or cutoff. As the following case histories indicate, this technique has been used in relatively shallow mine and limestone applications. The depth limit of its application has not been determined. An economic study would have to be made, comparing the larger diameter holes that would have to be drilled and cased through the cavity versus other techniques of forming grout columns.

Case Histories

CASE 1. In the late 1960s an incinerator foundation in Steelton, Pennsylvania was to be installed on spread footings. The original test boring missed a large cavity that was beneath some footings of the structure. The overburden and limestone bedrock was more than adequate to take the design loads of the footings, but the engineer was concerned with collapse of the solution cavity at some future date. After investigating the cost of flushing the cavity and pumping grout or concrete into the void, the engineer elected to support the cavity with a series of fabric-formed columns on a grid pattern. Five-inch-diameter holes were drilled to intercept the floor and tubes of woven nylon fabric were snaked through the drill hole into the cavity. They were then filled with concrete to their predetermined diameter of 2 feet. The contractor, engineer, and authority involved feel that the costs expended for forming these columns in the cavity were far less than any alternative system.

CASE 2. A golf course clubhouse in Latrobe, Pennsylvania was to be founded on a site where previous underground mining was known to have occurred.[13] Thus the existence of fissures, voids, and seams was very likely and of concern as to the foundation design. The area was stabilized by installing 16 Fabriform mortar piles to depths averaging 50 feet. Six-inch-diameter holes were drilled to bedrock or refusal. As shown in Figure 7.5, a small-diameter grout pipe inside a Fabriform closed-end sleeve was inserted in the drill hole to full depth. The sleeve

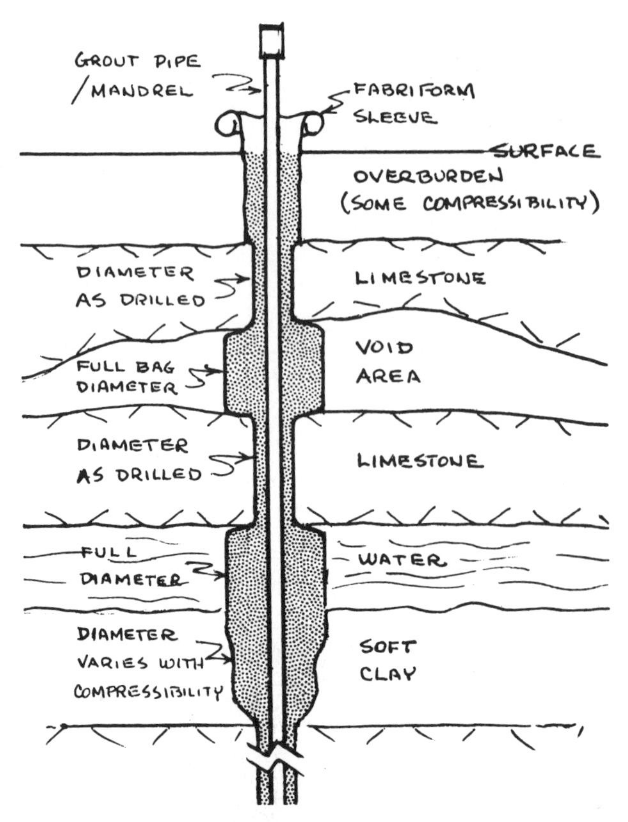

FIGURE 7.5. Idealized sketch of completed cross section of Fabriform® mortar piles, showing grout pipe within fabric sleeve placed in previously drilled 6-in. diameter hole (by B. A. Lamberton, Intrusion Prepakt, Cleveland, Ohio).

was then pressure injected with mortar as the grout pipe was gradually withdrawn. In zones of low lateral resistance, that is, voids, soft soils, and so on, the sleeve expands to densify the local area and/or to provide an end bearing, as shown in Figure 7.5. The sleeves can expand up to 16 to 24 inches. For additional load-carrying capacity the piles can also be reinforced.

7.4 Patent Bibliography for Use of Fabrics in Construction Forms for Concrete and Grout

Of all the areas where synthetic fabrics have been used in construction their use as a concrete form has resulted in the most patents; a partial list of these, with the authors' interpretations, is presented to guide the reader.

Patent Number	Issue Date	Inventor	Title and Brief Abstract
3,099,911 (reissue patent 25,614)	8/63	Turzillo	"Means of Grouting and Concreting—Flexible but porous fabrics used" as bags to contain grout or concrete
3,234,741	2/66	Ionides	"Layers or Screens for preventing or minimizing Fluid Flow Through Surfaces"—Among other disclosures covers an impermeable membrane to which tubes are attached and inflated with water or a slurry
3,344,609	10/67	Greiser	"Prevention of Beach Erosion and Encouragement of Land Restoration"—The bank is covered with a thin, flexible sheet material pervious to water but impervious to sand; the sheet is then held in place by inner engaged concrete blocks
3,345,824	10/67	Turzillo	"Methods and Means for Bracing or Bolstering Sub-Aqueous Structures"—Expands on patent 3,234,741 by providing for a support for the porous fabric bag
3,383,864	5/68	Turzillo	"Means of Protecting or Repairing Scoured Areas of a Situs"—By injection of a liquid grout into a flexible fabric bag, first to expand portions of the bag into a trench provided in the scoured area, and then to fill the remainder of the bag overlying the scoured area
3,396,542	7/68	Lamberton	"Methods and Arrangements for Protecting Shore Lines"—A pair of large sheets of flexible material at least in part porous are joined around their entire outer

Patent Number	Issue Date	Inventor	Title and Brief Abstract
			periphery; a cementitious slurry is injected into the space between the two sheets
3,425,227	2/69	Hillen	"Form for Constructing a Slab for Talus or Bottom Protection"—Similar to the above Lamberton patent with each point of attachment permits the passage of water whereby the hydrostatic pressure on both sides of the slab may be equalized
3,425,228	2/69	Lamberton	"Fabric Forms For Concrete Structures"—Similar to Patent 3,396,542 but adds a third sheet of flexible material
3,438,207	4/69	Turzillo	"Method of Making Concrete Retaining Wall in Earth Situs"—Provides a method of excavation where layers of fabric bags are inflated with cement mortar then an excavated cavity is extended beneath the bag and the process repeated
3,474,626	10/69	Colle	"Method and Means for Protecting Beaches"—Improves the Lamberton patent by adding a filter cloth to the exterior of the bottom fabric
3,486,341	12/69	Huesker-Stiewe et al.	"Forms for Concrete or the Like"—Describes a method of forming a mattress of concrete with the two fabric layers joined by ties
3,492,823	2/70	Lamberton	"Methods and Apparatus for Forming Elongated Hardened Concrete Bodies by Pressure-Grouting"—A flexible porous tube is positioned in any opening in the earth's surface; a grout is pumped and pressure maintained until the water cement ratio is reduced so that the grout is no longer flowable

Patent Number	Issue Date	Inventor	Title and Brief Abstract
3,520,142 3,524,320 3,570,254	7/70 8/70 3/71	Turzillo	"Method and Means for Protecting an Earth Situs Against Scour" —Adds anchor elements and internal ties between two fabric layers inflated with concrete
3,565,125	2/71	Hayes-Currier	"Dual Wall Fabric with Circular Connecting Points"—Describes a double layer fabric to be inflated with concrete and the spacing means woven into the fabric by the loom
3,786,640	1/74	Turzillo	"Means and Methods for Producing Stepped Concrete Slope Structures" —Laterally elongated bags are attached to a bottom mat and inflated with concrete
3,837,169	9/74	Lamberton	"Reinforced Mattress for Protecting Shore Lines and the Like"—Cords are used to connect the upper and lower fabric to control the amount of inflation by an injected cementatious slurry
3,871,182	3/75	Estruco	"Method of Protection for Slopes and Crests of Rivers Channels and the Like"—Permeable tubular casing are simultaneously filled with fresh concrete while being placed
3,984,989	10/76	Turzillo	"Means for Producing Sub-Aqueous and Other Cast-in-Place Concrete Structures In-Situ"—Uses an open work matrix to control expansion of fabric as it is inflated with fluid mortar

7.5 *References*

1. K. Terzaghi and Y. Lacroix, "Mission Dam: An Earth and Rockfill Dam on a Highly Compressible Foundation," *Geotechnique*, March 1964, pp. 13-50.

2. J. P. Welsh, "PennDOT Utilizes a New Method for Solving Scour Problems Beneath Bridge Structures," *Highway Focus,* Vol. 9, No. 1, May 1977, pp. 72-81.
3. G. den Hoedt and K. A. G. Mouw, "The Application of High Strength Woven Fabrics in Hydraulic Engineering Constructions," 7th International Congress Koninklyke Vlaamse Ingenieursvereniging V.Z.W., Vol. 2, May 1978, pp. 11-1, 9.
4. ———, "Six-Lane Tunnel Sits on Sacks of Grout," *Eng. News-Record,* Nov. 9, 1967, pp. 104-105.
5. J. P. Welsh, "Utilization of Synthetic Fabrics as Concrete Forms," in Proceedings of the International Symposium on New Horizons in Construction Materials, Lehigh University, Bethlehem, Pa., Nov. 1-3, 1976.
6. M. Kupfer, "Engineering Data on Pile-Renu Process," Ocean Operations Division, Dillingham Corp., San Diego, Ca., April 1969.
7. J. P. Welsh, *"Fabriform Pile Jacketing System,"* in *Design and Installation of Pile Foundations and Cellular Structures,* Envo, Lehigh Valley, Pa., 1970.
8. J. P. Welsh and R. M. Koerner, "Innovative Uses of Synthetic Fabrics in Coastal Construction," in Proceedings of Coastal Structures '79, March 14-16, 1979, Washington, D.C., pp. 364-372.
9. B. A. Lamberton, "Fabric Forms for Concrete: A New Technology with a Long History," CExpo '77, San Francisco, Cal., Oct. 18, 1977.
10. Product Literature, Frankipile, Ltd., Davis House, High Street, Croydon CR 9 1PN, England.
11. R. M. Koerner and C. Mukhopadhyay, "The Behavior of Negative Skin Friction on Model Piles in Medium Plasticity Silt," *Highway Res. Rec.,* Washington, D.C., No. 405, Oct. 1972, pp. 34-44.
12. R. E. Gray, J. C. Gamble, R. J. McLaren, and D. J. Rogers, *State of the Art of Subsidence Control,* Report No. ARC-73-111-2550, Appalachian Region Comm. and Dept. of Environmental Resources of Pa., Washington, D.C., Dec. 1974.
13. Product Literature, Intrusion Prepakt, Inc., 1705 The Superior Building, Cleveland, Ohio 44114.

8

Impermeable Fabrics

Once a construction fabric is made nonpervious an entirely new spectrum of construction uses emerges. There uses can be categorized in the following areas:

- Air-supported and tension structures.
- Water-filled structures.
- Self-sustaining structures.

Such applications can occur above ground, below ground, or in water.

The field is seen by many to be a major market for new structures and fabrics because of the infinite variety of geometric configurations that can be formed, their relatively low cost, and relatively rapid construction times (see Figure 8.1 for an example). However, there are cautions that must be considered. These are vunerability to vandalism, deterioration by the environment, and decay of structural properties. With time and experience in the design, manufacture, and construction of the fabrics most, hopefully all, of these disadvantages will be eliminated.

Of the three areas presented in this chapter, Air-supported and tension structures currently have received wide acceptance and are the major topic of interest covered. It should also be recognized throughout this chapter that this topic boarders closely on various types of impermeable liners as used in reservoirs, tanks, and pollution control facilities.[1]

8.1 Air-Supported and Tension Structures

Background

The first practical application of air-supported structures in the United States was in the late 1940s for use as radomes covering large radar antennas.

FIGURE 8.1. Sample use of an impermeable fabric enclosure.

These structures were approximately 40 feet high and 50 feet in diameter. Originally made from neoprene-coated fiberglass material, the need was filled for a thin nonmetallic weather-resistant enclosure. Subsequent materials used were neoprene-coated nylon, Fortisan rayon, and dacron materials.[2]

Today, air structures and their related tension structures range in size from small food service pavillions to a ten-acre roof for an 80,000-seat athletic stadium. In general, air structures are accepted because of their low cost, quick erection, free span space, and seasonal adaptability. Tension structures incorporate modern geometric designs for a wide variety of uses. Table 8.1 shows their current market statistics.[3] In the table, about 15 percent of the sales are for tension structures, which are predominantly vinyl-coated polyester. Growth in this area is anticipated to be very high.

Applications currently fall into the following categories, as supplied by the Canvas Products Association International (CPAI).[4]

- Industrial warehousing—to satisfy emergency, seasonal, or continued demand for industrial warehouse space.
- On-site storage shelter—to provide on-site weather protection for products of materials that may require storage in the field.

TABLE 8.1. **Reported Air Structure and Tension Structure Sales in 1975 and 1976**

Use	1975 Dollar Volume	Percentage of Total	1976 Dollar Volume	Percentage of Total
Industrial	4,150,000	30.2	7,000,000	36.3
Recreational	3,800,000	27.7	4,525,000	23.5
Government and military	3,315,000	24.1	4,250,000	22.0
Large-span structures	1,620,000	11.8	1,950,000	10.0
Exhibits and specialties	855,000	6.2	1,575,000	8.2
Totals	$13,740,000	100%	$19,300,000	100%

Source: Ref. 3.

- Construction shelter—to provide shelter and protection for continuing construction during inclement weather.
- Recreational enclosures—to provide enclosed recreational space for swimming pools, indoor tennis, ice hockey, and all-purpose recreation for schools and municipalities.
- Social activity shelters—to provide weather protection for special exhibits, church services, traveling shows, and so on.
- Agricultural enclosures—to provide environmental control weather protection for year-round crop growth.
- Manufacturing facilities—to provide enclosed manufacturing space for operations that require large, clear span work areas.
- Disaster shelters—to provide timely sheltered space as emergency hospitals, schools, field centers, and so on, in the event of a public disaster.
- Ecological enclosures—to provide ecological and environmental control for water and sewage treatment or other similar applications.

Concepts and Design

An air-supported structure has three basic components: a fabric envelope, an inflation system, and an anchoring system. Regarding the fabrics currently in use, Table 8.2 show that PVC-coated polyesters account for over one-half of the total market.[3]

The inflation system includes a primary and an auxiliary blower, the latter being necessary during times of power failure and extremely high winds. Anchorage systems consist of continuous surface anchoring methods

TABLE 8.2. Air and Tension Structure Fabric Usage

Fabric	1975 Square Yards	Percentage of Total	1976 Square Yards	Percentage of Total
PVC-coated polyester	387,000	53.3	405,000	53.3
PVC-coated nylon	193,000	26.5	202,000	26.5
Neoprene-coated nylon	97,000	13.5	102,000	13.5
Teflon-coated fiberglas	36,000	5.0	38,000	5.0
Unsupported vinyl film	12,000	1.7	13,000	1.7
Totals	725,000	100%	760,000	100%

Source: Ref. 3.

or internal separated embedded anchors. The entire system is shown schematically in Figure 8.2. Shapes are limited to spherical, cylindrical, or combined and are somewhat dictated by the anchor layout system. There are obvious limitations in this regard so as not to cause high stress concentrations.

CPAI, on the basis of an inflation pressure of 1.5 inches of water pressure (approximately 0.058 pounds per square inch), a design wind load of 75 miles per hour, and a safety factor of 4 (based on strip tensile test), has developed guidelines (Table 8.3) for minimum values of material strength. The problem of fabric strength and behavior, however, is a very complex one. Skelton[5] brings out the fact that the fabric in an inflated structure does not function in uniaxial tension but in a biaxial stress mode; for example, in a spherical structure the stresses in two perpendicular directions are equal and in a cylindrical structure the ratio is approximately

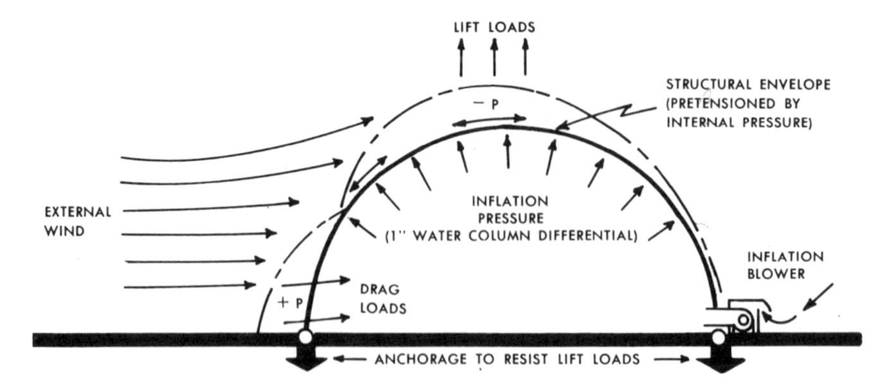

FIGURE 8.2. Aerodynamic and inflation loading (from Ref. 4).

TABLE 8.3. Minimum Design Criteria for Material Strength

Size (ft)		Tensile strength		Tear strength		Dead load		Coating Adhesion, Minimum
Width	Height	Strip (lb/in.)	Grab (lb)	Trapez. (lb)	Tongue (lb)	Room Temperature	160°F	(lb/in.)
30	15	112	160	17	32	56	28	
40	20	142	202	20	38	72	36	
50	25	175	250	23	42	88	44	
60	30	210	300	26	50	106	53	
70	30	245	350	30	58	122	61	
80	35	280	400	33	65	140	70	10 lb/in.
90	40	312	445	36	70	156	78	
100	40	350	500	40	78	175	88	
110	40	375	536	44	86	188	94	
120	42	420	600	48	100	210	105	
130	45	470	670	55	110	235	118	
140	50	490	700	62	122	245	122	
150	60	532	760	70	138	266	133	

Source: Ref. 4.

2:1. For a 16 ounce per square yard coated fabric used in inflated structures Skelton presents the load versus elongation information of Figure 8.3. The data are for uniaxial (strip) and biaxial (1:1 load ratio) tests along the warp and filling directions. Note the extremely high modulus for the warp direction in biaxial load and the difference in behavior between strip tests and biaxial tests in the fill direction. Consequences of this type of behavior can be elongation mismatch, puckering of fabric, and reduced seam efficiency.

Joint strength is critical, with the usual joining being accomplished by heat sealing, cementing, or sewing. All joint seams must be capable of developing the full strength of the envelope material. This is usually measured on the basis of the strip tensile test. See Figure 8.4 for the common types of seams.

Air-supported structures are largely flame resistant owing to the existing pressure difference between the inside and outside of the fabric enclosure. If a localized burnout occurs, the air flow outward is so great that combustion cannot be supported. No air-supported structure has ever been destroyed by fire.[4] This does not mean that fire-resistant qualities are not desired. Materials must meet local codes and are particularly important in tension structures, which do not have the benefit of a high internal pressure to extinguish a potential fire.

The inflation system is used to inflate and shape the fabric envelope. It is

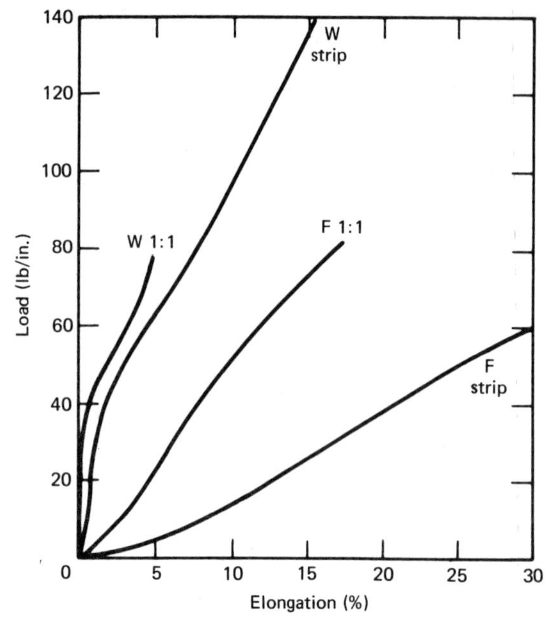

FIGURE 8.3. Uniaxial and Biaxial tensile behavior of coated fabric (from Ref. 5).

the stabilizing factor in air-supported structures. The air pressure required is for design wind and snow conditions and must have an air flow velocity to compensate for air leakage and yet maintain the required operating pressure. CPAI recommends an inflation pressure of about 1.5 inches of water for 75-mile-per-hour winds,[4] which is equivalent to 0.058 pounds per square inch or 8.3 pounds per square foot above atmospheric pressure. The capacity of the equipment is obviously dictated by the size and tightness of the envelope, the type and number of doors, and requirements for air venting.

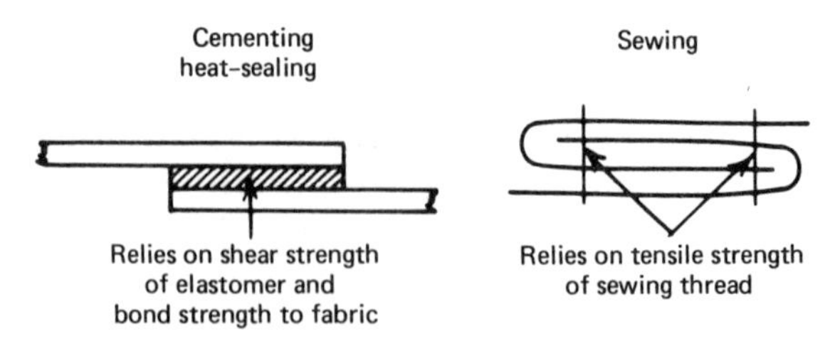

FIGURE 8.4. Factors controlling the strength of various seams (from Ref. 5).

The anchorage system attaches the envelope to the ground as well as providing for a seal of the envelope at its base. It must be designed to resist both lift forces developed by the envelope and aerodynamic wind forces. Some typical anchorage systems are shown in Figure 8.5. The anchorage forces mobilized are quite high, as Table 8.4 indicates.

Case Histories

As reflected in the prior discussion, case histories of air-supported and tension structures are numerous, with new uses being constantly deve-

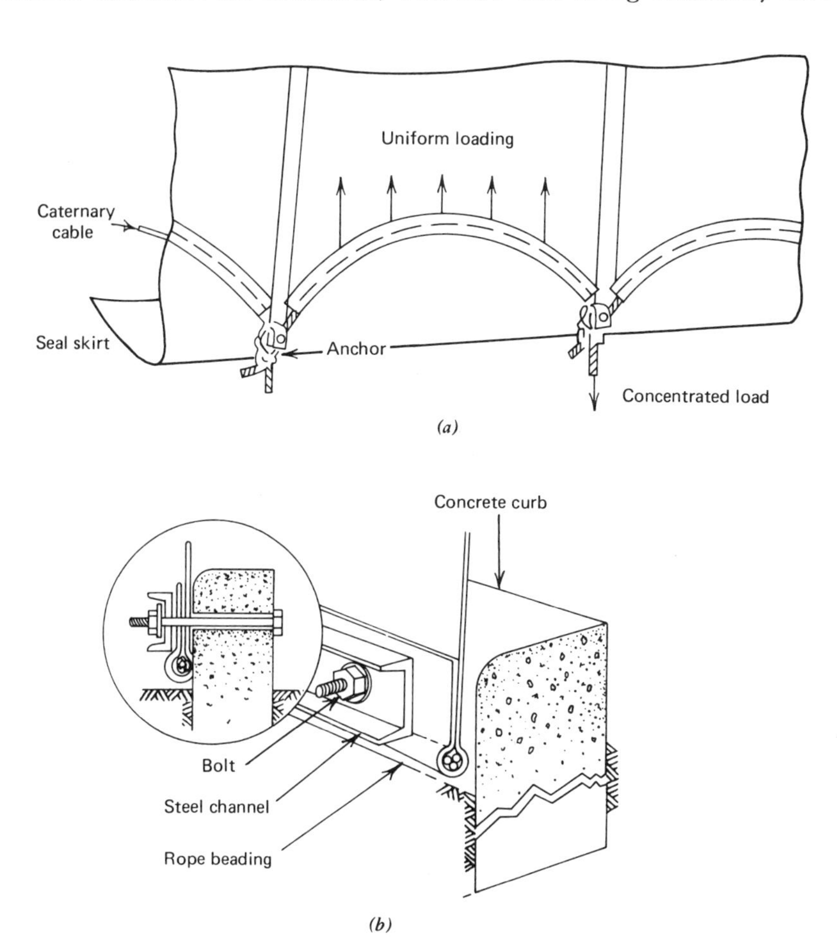

FIGURE 8.5. Types of air-supported anchorage systems: (*a*) intermittent catenary base attachment; (*b*) continuous rope edge attachment, for use with concrete curb anchorage (from Ref. 4).

TABLE 8.4. Design Anchor Loads[a]

Structure Size (ft)		Anchor Load (lb/ft) at 75 mph[b]
Width	Height	
30	15	660
40	20	880
50	25	1,100
60	30	1,320
70	30	1,500
80	35	1,780
90	40	2,000
100	40	2,250
110	40	2,500
120	42	2,650
130	45	2,850
140	50	3,000
150	60	3,200

Source: Ref. 4.

[a] Maximum design anchor loads are based upon 75-mph wind and 1.2-weight of gravity (wg) internal static pressure with a safety factor of 2. For unreliable soil conditions, use a safety factor of 3 or more. Values are based upon vertical panels and cylindrical (higher load) shape. Anchorage for unique shape designs requires special engineering consideration.

[b] Anchorage design and site subsoil conditions should be verified by analytical or test data.

loped. As seen in Figures 8.6 through 8.13, the range of applications is varied and depends greatly on the innovativeness of the designer.

CASE 1. Figures 8.6 and 8.7 show an artist's rendering and the inside of the completed air-supported structure that forms the roof of the 80,000-seat Pontiac Stadium in Pontiac, Michigan. The dome fabricator was Birdair Structures, Inc., of Buffalo, New York.[6] The roof was installed by the Contracting Division, Owens-Corning Fiberglas Corporation of Toledo, Ohio. The dome is restrained and shaped by a network of 18 steel cables, which vary in length from 550 to 750 feet. A concrete and steel compression ring encircles the top of the stadium and anchors the cables. The dome was inflated to height of 205 feet above the field by 29 large fans in about 2 hours. Once inflated, only 2 or 3 fans are required to supply the normal operating

FIGURE 8.6. Artist's rendering of Pontiac Stadium showing air-supported dome with cable restraining system (photo by O'Dell, Hewlett, and Luckenback, Inc., Birmingham, Mich.).

FIGURE 8.7. Inside of Pontiac Stadium, showing air-supported structure dome roof (photo provided by Birdair Structures, Inc., Buffalo, N.Y.).

air pressure of 3.5 to 4.0 pounds per square foot. The fabric used for this project was a Teflon-coated fiberglas. It is believed to be the largest fabric enclosure (approximately 10 acres) constructed to date.

CASE 2. Figure 8.8 shows photographs of a recreational facility owned by the State University of New York at Buffalo, and constructed by Birdair

FIGURE 8.8. Recreational facility at the State University of New York at Buffalo (photo by Birdair Structures, Inc., Buffalo, N.Y.).

Structures, Inc.[6] This air-supported structure is 120 feet wide by 262 feet long by 42 feet high (maximum) and encloses a floor area of about 31,000 square feet. The fabric is a vinyl-coated nylon and incorporates an inner liner attached to the outer fabric. This provides for an insulating air space between the two materials. The anchorage system consists of $\frac{3}{4}$ inch steel cables at 16-foot spacings. The inflation system operates at 1.25 inches of water pressure.

CASE 3. The University of Riyadm recreational facility in Saudi Arabia required a low silhouette due to wind forces and sand storms and is shown in Figure 8.9. The structure consists of an air-supported fabric which is cable reinforced and supported within earth berms around the perimeter. The roof area is 44 by 133 meters and consists of fourteen panels supported by thirteen $1\frac{1}{2}$-inch-diameter steel cables. Design, fabrication, and installation was by Birdair Structures, Inc.[6] The outer membrane is a fiberglas fabric coated with Teflon. A second, and inner, membrane of similar material but lighter and somewhat porous was used to improve the thermal and acoustical properties. Located within the earth berm are mechanical rooms, locker rooms, offices and other facilities.

CASE 4. Figure 8.10 shows an example of a tension-supported structure used as a student center at LaVerne College in La Verne, California. The tent-like structures cover 1.4 acres and consist of a series of conical shapes supported by four columns. One inch diameter steel cables extend radially

FIGURE 8.9. Artist's rendering of recreation facility for University of Riyadm, Saudi Arabia (photo by Building Sciences, Inc., Towson, Maryland).

FIGURE 8.10. Tension structures of LaVerne College Student Center (photos by Birdair Structures, Inc., Buffalo, N.Y.).

from the top of the columns to a curvalinear concrete compression ring close to the ground. The roof membrane was patterned so that it could be pretensioned by jacking up the column to introduce a predetermined level of tension into the radial cables. The skin is a fiberglas fabric coated with a special formulation of Teflon. The roof membrane was fabricated and installed by Birdair Structures, Inc.[6]

Most of the previous examples of air-supported and tension structures utilized a woven fiberglas fabric treated with a Teflon resin. Some technical properties of an architectural grade of Teflon-coated fiberglas are given in Table 8.5.

CASE 5. A somewhat different concept in tension structures is offered by Seaman Building Systems,[7] with their patented Portomod® and Tension Span™ Structures. As seen in Figures 8.11 through 8.13, these structures consist of a unitized modular clear span curvilinear structural frame over which a flexible membrane is prestressed. Geometrically and in plan view the structures form a polygon of an equal number of sides or a polygon

TABLE 8.5. Teflon-Coated Fiberglas Properties

	Sheerfill I	Sheerfill II	Sheerfill III	Method or Standard
Coated fabric weight, average (oz/yd^2)	45	37.5	13.5	Method 5041[a]
Strip tensile strength (lb/in.)				
Warp, dry	800	520	400	Method 5102
Fill, dry	700	430	320	
Warp, wet	700	440	340	
Fill, wet	600	360	270	
Trapezoidal tear (lb)				
Warp	60	35	24	Method 5136[a]
Fill	80	38	22	
Flame resistance				
After flame (sec)	1	1	1	Method 5903[a]
After glow (sec)	1	1	1	
Char length (in.)	0.25	0.25	0.25	
Translucency, 100% bleached	8–12	10–15	20–25	ASTM E-424-71

Source: Ref. 6.
[a]Federal test number 191-GSA.

FIGURE 8.11. Outside of Dolphin Show Theater, Houston, Texas (photo by Seaman Building Systems, Sarasota, Fla.).

halved and separated by any number of parallel bays. Current standard clear span widths are 55, 72, 90, 110 and 124 feet. Custom designed structures include clear span widths up to 200 feet. The base fabric is a woven Dacron polyester sandwiched between vinyl compound coatings and a top coating. The structural steel space frame is bolted together at the job site and designed to meet specific load requirements such as wind, snow, vibration, and so on.

8.2 Water-Filled Structures

Background

While nowhere near as significant or common as air-supported or tension structures, it is possible to use water as a fill for impermeable fabrics. The resulting shape takes the form of the fabric and is supported by hydrostatic pressure, which varies linearly with increasing depth. Overpressures can also be used with comparably stronger fabrics.

FIGURE 8.12. Inside of Dolphin Show Theater, Houston, Texas (photo by Seaman Building Systems, Sarasota, Fla.).

Key elements in water-filled structures are the membrane itself, the anchorage system, the pumping system, and an adequate supply of water.

Concepts and Design

There is a definite lack of information on actual design data for water-filled structures. To our knowledge, only hypothetical plans for large projects and a few small projects have been performed. All are concerned with temporary, that is, intermittent, damming of water in rivers or harbors. Some of these applications follow.

Case Histories

CASE 1. By far the most ambitious use of water-filled structures made from fabrics is the damming of the Adriatic Sea at entrances to the City of Venice. Conceived by Colamussi of the University of Bologna and furthered by Pirelli of Milan, the construction firm of Fulanis and the synthetic fiber group of Enka devised three giant tubes to span the three lagoon entrances to Venice.[8] Normally the dams lie flat on the sea bed, allowing sea traffic to pass above them. When adverse weather conditions are expected, six pump-

FIGURE 8.13. Inside of tennis enclosure at Racquet Center at Cooper Springs, Meyersville, N.J. (124 by 267 by 47 feet) (photo by Seaman Building Systems, Sarasota, Fla.).

ing stations pump seawater into the dams until they reach the necessary height (see Figure 8.14). The pumping should be completed in 30 to 120 minutes, which is ample time considering that a 6-hour warning of high seas is customary. The fabric dams are anchored to concrete pylons with steel cables. Project data are given in Table 8.6. It appears as though the

Flexible dam in unfilled condition; Ocean-going ships can pass freely through the opening into the lagoon and harbor.

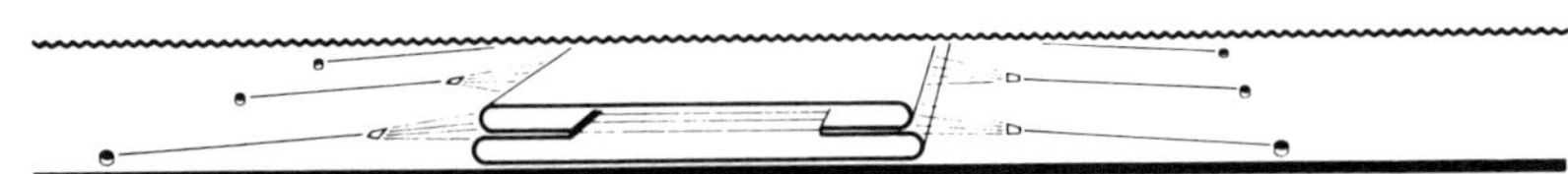

The sinkable high-water protection dam is designed as 30m wide and 15m high (when filled). It will be 470m long for the Malamocco Canal, 550m long for the Chioggia Canal and 900m long for the Lido Canal.

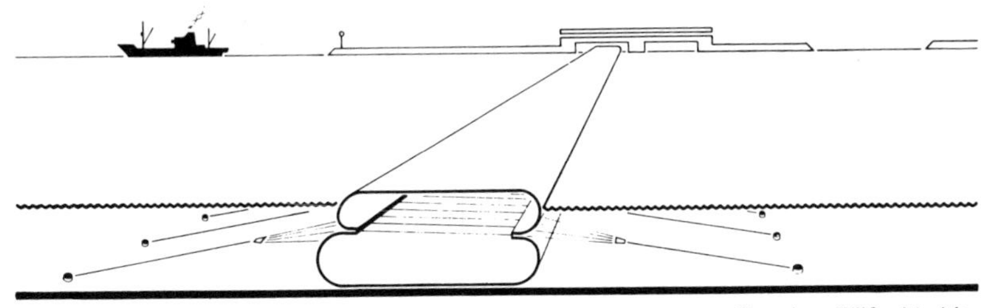

At flood stage the dam is pumped to its maximum height. Sketches; Wilfred Lubitz

FIGURE 8.14. Proposed scheme of damming off channel leading to Venice using inflatable fabric tubes filled with seawater (from Ref. 8).

TABLE 8.6. Data on Required Water-Filled Tubes for Channel Dams Entering Venice

Description	Lido Channel	Malamocco Channel	Chioggia Channel
Width (m)	900	470	550
Average depth below mean sea level (m)	9.2	14.0	8.2
Maximum depth below mean sea level (m)	15.0	19.4	10.8
Separating area (m^2)	8,300	6,600	4,500
Surface area of dams (m^2)	73,400	50,200	36,300
Filling volume (m^3)	285,500	278,800	119,900
Number of pylons	80	68	52

Source: Ref. 8.

FIGURE 8.15. Po River Valley tests of inflatable fabric tubes to seal off river channel (from Ref. 8).

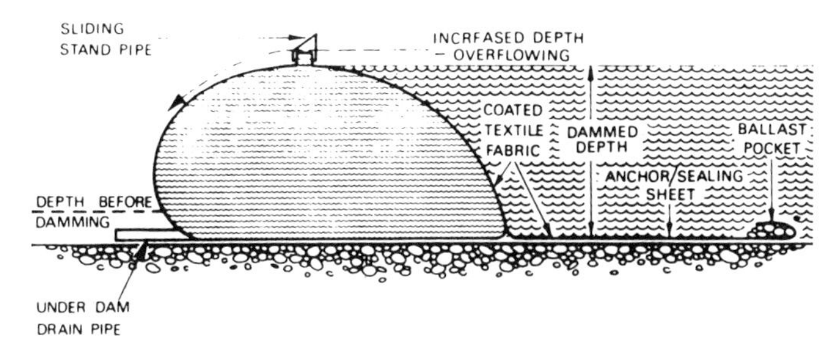

FIGURE 8.16. Schematic diagram of Conenco's Tubedam (from Ref. 9).

fabric will be a neoprene-coated polyamide made from high-tenacity nylon yarn.

Figure 8.15 shows photographs of the deflated and partially inflated tests in the Po River Delta.

CASE 2. Tubedam is an inflated tube barrier developed by Conenco International, Ltd., of Canada. It is held in place and sealed by hydrostatic pressure, is quickly installed and removed, and is particularly useful for increasing the levels of small streams and rivers liable to large or rapid variations in flow. The system is designed that when the overflow reaches a predetermined level, the tube deflates automatically. Figure 8.16 shows a schematic of this system.

8.3 *Self-Sustaining Structures*

Background

Once a fabric is coated and rendered impervious it can develop sufficient stiffness to become self-sustaining, or at least partially self-sustaining. Such structures are of necessity relatively small and/or of small span length. Many tension structures described in Section 8.1 fall into this category. In general, the structures are of a temporary nature, but as several of the case histories following show, they can be of a quasi-permanent nature.

Examples

Portadam® is a patented cofferdam by Conenco International, Ltd., of Canada, and can support hydrostatic pressures of up to 10 feet (see Figure 8.17). It consists of a braced metal support system carrying a reinforced

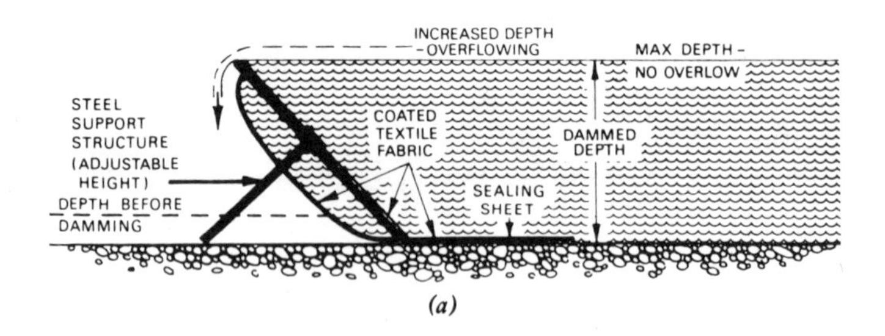

FIGURE 8.17. Schematic diagram (a) and photographs (b and c) of Conenco's Portadam® system. Photographs courtesy of Conenco International, Ltd., Canada.

fabric membrane which is sealed at its base and sides. Various coated fabrics are available to resist oil, acids, extreme temperatures, and so on. Where the ground or supporting foundation is relatively flat a number of uses are possible. These are as follows:

- Flood protection.
- Cofferdam.
- Underwater repairs.
- Underwater construction.
- Raising water levels.
- Oil spill and industrial waste containment.
- Natural swimming pools.
- Cleaning lakes, rivers, and ponds.
- Environmental.
- Ecological.
- Overflow back-up system.
- Launching ramp construction.
- Water storage.

8.4 References

1. W. B. Kays, in *Construction of Linings for Reservoirs, Tanks, and Pollution Control Facilities,* M. D. Morris, ed., Wiley, 1977.
2. W. Bird, "History of Air-Structures in USA," in *Proceedings of the International Conference on Practical Applications for Air Supported Structures, Oct. 28-29, 1974, Los Vegas, Nevada,* CPAI, St. Paul, Minn., pp. 55-73.
3. ______ "Air Structures and Tension Structures Grow Dramatically," *Indust. Fabric Prod. Rev.* Vol. 54, No. 3, July 1977, pp. 42-46.
4. ______, "Minimum Performance Standard for Single-Wall Air-Supported Structures," March 31, 1971, CPAI, St. Paul, Minn.
5. J. Skelton, "Comparison and Selection of Materials for Air-Supported Structures," *J. Coated Fibrous Mater.*, Vol. 1, April 1972, pp. 208-221.
6. Product Specification Record, Birdair Structures, Inc., Buffalo, N.Y. 14225.
7. Seaman Building Systems, 2028 E. Whitfield Blvd., Sarasota, Fl. 33580.
8. Information Sheets from Enka Glanzstoff AG, Wuppertal, Germany.
9. Product Literature, Conenco International, Ltd., 39 Esna Park Dr., Markham, Ontario, L3R 1C9 Canada.

9

Guidelines and Current Research and Development

9.1 Guidelines for Fabric Use

McGown,[1] in an excellent overview of the current fabric use situation, feels it is doubtful if a general quantitative specification of fabric properties can be prepared. He further states that field experience has been the only guide for selection of fabrics, thus the extent of their proven satisfactory performance is vitally important. Indeed, we agree with these observations at least for the present time. This book, which is essentially a compilation of field experiences, is an attempt at satisfying this need. However, as far as most public works are concerned, this status is unacceptable. Quantification of fabric properties, that is, as a logical, systematic, and technically accurate methodology for design and competitive bidding, is necessary. In this regard, many attempts are being made by manufacturers, users, research organizations, and universities.

To satisfy the current need for fabric use guidelines (or specifications) attempts have been made and published in the open literature. In this section we have collected in an abbreviated form some of these documents. Reference 2 should also be consulted since it is a more complete compilation of the same specifications for fabrics from the states of Alabama, California, Illinois, and New York and from the U.S. Army Corps of Engineers and the USDA Forest Service than we have presented here.

9.1.1 Corps of Engineers Guidelines for Subdrains

Extending the work of Calhoun[3] to deal with both woven and nonwoven fabrics for use in an aggregate-wrapped fabric drainage system, the Corps of Engineers[4] recommends the following fabric properties for use in underdrain systems.

- For fabrics adjacent to granular soils containing 50 percent or less by weight of minus number 200 sieve size particles, it is required that

$$\frac{\text{85\% size of soil}}{\text{opening size of EOS sieve}} \geqslant 1$$

 where EOS is the equivalent opening size as described in Section 2.3.3. Furthermore, the total open area of the fabric should not exceed 36 percent.
- For fabrics adjacent to soils finer than above, the EOS should be no larger than the opening in the U.S. Standard Sieve number 70 (0.0083 inch) and the open area of the fabric cannot exceed 10 percent.
- To reduce the chance of clogging of the fabric during its performance lifetime, no fabric should have an EOS smaller than the opening of a U.S. Standard Sieve number 100 (0.0059 inch) or an open area less than 4 percent.

9.1.2 USDA Forest Service Guidelines and Specifications

A leader in the use of fabrics, particularly in construction of low-volume roads (often over very unstable soils) and of retaining walls, the U.S. Forest Service has prepared guidelines and specifications for various construction uses (see Steward et al.[5]). Table 9.1 is a sample guideline for nonwoven fabrics as compiled in reference 2. All values listed are the minimum requirements for the fabric property listed. The table is categorized in four groups according to construction use. Note that these are special guidelines used on individual projects in Region 8 (Southern Region) of the U.S. Forest Service. The only fabric specification required for use by the U.S. Forest Service is Section 720, "Plastic Filter Cloth," which is for *woven* plastic filter fabrics. It is available from the U.S. Government Printing Office, Washington, D.C.

United States Dept. of Agriculture Forest Service Specification

Section 720 Plastic Filter Cloth

720.01 GENERAL. Plastic filter cloth shall be a pervious sheet of plastic monofilament yarn woven into a uniform pattern with distinct and measurable open-

TABLE 9.1. Sample Fabric Guidelines Used by Region 8 (Southern District) of U.S. Forest Service

	Type of Construction Use			
Fabric Property	Brush Barriers and Silt Fencing	Roadway Reinforcement	Gravel-Filled Underdrains	Rip-Rap Liner
Fabric type	Nonwoven	Nonwoven	Nonwoven	Nonwoven or woven
Water permeability [(gal/min)/ft^2]	500	400	350	350
Thickness (mils)[a]	70	70	90	100
Grab strength (lb)[b]	–	100	90	200
Grab elongation (%)[b]			15 minimum 20 maximum	
Burst strength (psi)[c]	200	–	250	400
Elongation at 40% (%)			50 in 24 hr	
Trapezoidal tear (lbs)[d]	50	50	–	90
Uv degradation (% min. yearly loss)	25	–	–	–
EOS	70 minimum 100 maximum	–	–	–
Siphonage (qts/hr for 2-in. fabric)	–	0.3	0.3	0.3

[a] ASTM test D-1777.
[b] ASTM test D-1682.
[c] ASTM test D-231.
[d] ASTM test D-2263.
Source: Ref. 2.

ings. The plastic yarn shall be synthetic polymer composed of at least 85 percent by weight of propylene, ethylene, or vinylidene-chloride, and shall contain stabilizers and/or inhibitors added to the base plastic to make the filaments resistant to deterioration due to ultraviolet and/or heat exposure. The cloth shall be calendered so that the yarns will retain their relative position with respect to each other. Edges of the cloth shall be selvaged to prevent the outer yarn from pulling away from the cloth. Cloth shall be free of defects, rips, holes, or flaws.

The Contractor shall furnish a mill certificate or affidavit signed by an official from the company manufacturing the cloth. The mill certificate or affidavit shall attest that the cloth meets the requirements stated in these Specifications.

720.02 ACCEPTANCE TESTING FOR PLASTIC FILTER CLOTH. All brands of plastic filter cloth to be used shall meet the requirements of this Specification when tested according to the procedures contained in the U.S. Army Corps of Engineers document entitled, *Guide Specifications Plastic Filter Cloth*, No. CE 1310, May 1973.

TABLE 1. Cloths Tested by Corps of Engineers

Manufacturer or Fabricator	Trade Name	E.O.S. sieve No.*	Percent Open Area	Abrasion Resistance**
Carthage Mills, Inc. Erosion Control Div. Cincinnati, OH 45216	Filter X	100	4.6	Low
	Poly-Filter X	70	5.2	High
	Poly-Filter GB	40	24.4	High
Advance Construction Specialties Co. 1050 Texas Street Memphis, TN 38106	Erosion Control Fabric (Type I)	100	4.3	High
Erco Systems, Inc. P.O. Box 4133 New Orleans, LA 70118	Nicolon 66411	30	36.0	Low

*E.O.S. is "equivalent opening size," and is defined as the number of the U.S. Standard Sieve having openings closest in size to the filter cloth opening.
**For "high" abrasion resistance, the strength loss after testing shall not exceed 70 percent and the abraded strengths must be no less than 100 lb in the stronger principal direction and 55 lb in the weaker principal direction.

720.03 CLOTHS TESTED BY CORPS OF ENGINEERS. The plastic filter cloths of Table 1 have been tested by Corps of Engineers and found to meet specification requirements for the cloth types listed in Table 3, Subsection 720.05.

TABLE 2. Minimum Physical Requirements for Plastic Filter Cloth

Test	Minimum Strength % of Unaged Tensile Strength
Alkali treatment	90
Acid treatment	90
Low temperature treatment	85
High temperature treatment	80
Oxygen pressure	90
Freeze thaw	90
Weatherometer	65
	Test result
Brittleness	No failures at −60°F
Weight change in water	Less than 1.0%

TABLE 3. Minimum Unaged Strength Requirements for Plastic Filter Cloth

Cloth Type	Pretested Cloths	Stronger Principal Direction (tensile, lb)	Weaker Principal Direction (tensile, lb)	Burst (PSI)	Puncture (lb)	Seam Breaking (lb)
AB	Poly-Filter X Poly-Filter GB Erosion control fabric	200	200	510	125	195
C	Nicolon 66411 Filter X	180	100	250	65	90

720.04 CLOTH NOT PREVIOUSLY TESTED. If the Contractor elects to use a filter cloth other than listed in Subsection 720.03, he shall furnish test results performed as prescribed in Subsection 720.02. Test results shall be furnished at least 60 days prior to installation.

720.05 PHYSICAL AND STRENGTH REQUIREMENTS. Plastic filter cloth shall meet the physical requirements listed in Table 2 and the strength requirements listed in Table 3. Unless otherwise shown on drawings, Type AB filter cloths listed in Table 3 shall be used.

720.06 SECURING PINS. Pins for securing the cloth shall be of steel, a minimum of 3/16 inch in diameter, and at least 15 inches in length. Other equivalent securing devices may be substituted if recommended by the manufacturer.

9.1.3 State of Alabama Specifications

By far the most aggressive state in the use of construction fabrics is Alabama. This is particularly the case for fabric trench and French drain underdrain systems (e.g., see Lockett[6] which contains specifications and complete design problems) and also in the areas of erosion control and reinforcement using fabrics. Table 9.2 summarizes Alabama's specifications according to construction use (four categories) and is adapted from Ref. 2. It contains separate specifications for both woven and nonwoven fabrics.

9.1.4 Other State Department of Transportation Fabric Specifications

California has been active in the use of nonwoven fabrics for subdrains and uses the information of Table 9.3 as its specification. Illinois has a

TABLE 9.2. Sample Fabric Specifications of State of Alabama DOT Highway Department

Fabric Property	Type of Construction Use						
	Underdrains		Aggregate Liner		Rip-Rap Liner	Reinforcement	
Fabric type	Woven	Nonwoven	Woven	Nonwoven	Woven	Woven	Nonwoven
Grab strength (lb)[a]	200	90	200	90	200	200	90
Grab elongation (%)[a]	—	50	—	50	—	10–35	50
Burst strength (psi)[b]	500	—	500	—	500	500	—
Puncture strength (lb)[b]	120	—	120	—	120	120	—
Abrasion resistance (lb)[c]	55	—	55	—	55	55	—
Seam breaking strength (lb)[d]	180	—	180	—	180	180	—
Permeability[e]							
Minimum (cm/sec)	0.02	0.02	0.02	0.02	0.02	0.033	0.02
Maximum (cm/sec)	0.3	0.3	0.3	0.3	0.3	0.038	0.3
Coefficient of friction (Fabric/aggregate)	—	—	0.4	0.4	—	—	—
Granular soils (50% > no. 200)							
d_{85}/EOS, minimum (%)					4		
Open area, maximum (%)					35		
Fine soils (50% < no. 200)							
EOS					70–100		
Open area (%)					4–10		

Source: Ref. 2.

[a] ASTM test D-1682.

[b] ASTM test D-751.

[c] ASTM tests D-1175 and D-1682.

[d] ASTM test D-1683.

[e] AHD Permeability Test for Filter Fabric.

TABLE 9.3. California DOT, Division of Highways Specification

Property	Specification
Weight, ASTM D-1910 (oz/yd^2)	4.0 minimum
Thickness, ASTM D-1777 (mils)	50 maximum
Tensile strength, ASTM D-1682 (lb)	100 minimum
Elongation, ASTM D-1682 (%)	50 minimum
Toughness (strength × elongation)	6000 minimum

Source: Ref. 2.

similar specification for fabric underdrain systems (either woven or non-woven) with the minimum properties given in Table 9.4.

A very different approach than Illinois is taken by New York State DOT, which feels that specifications based on permeability and EOS vis-à-vis soil type is not adequate to design fabrics for drainage or erosion control practices. They preaccept fabrics for a specific application based on

- Sufficient strength for handling and placement.
- Adequate resistance to ultraviolet light, chemicals, and rot.
- Adequate EOS.
- Adequate permeability.
- Sufficient retention of suspended soil under free water flow.
- Fabric suitability for the specific use intended based on visual inspection, and model and/or field tests

If unique applications are encountered, special tests are devised, and New York has developed a number of new and revised tests for their use, for

TABLE 9.4. Illinois DOT Fabric Specification

Property	Specification
Weight, ASTM D1910 (oz/yd^2)	4.0 minimum
Grab strength, ASTM D1682 (lb)	100 minimum
Grab elongation, ASTM D1682 (%)	60 minimum
Fabric to retain soil $>$ 106 μm (no. 140 sieve)	
Fabric to pass soil $<$ 25 μm	

Source: Ref. 2.

example, the flow capacity test, puncture test, grab test, and weatherometer test.

Under this framework, New York has developed three sample descriptive specifications for underdrain, undercut and bedding applications, (see Ref. 2). New York State DOT has selectively approved many of the fabrics listed in Table 2.4.

The Commonwealth of Pennsylvania appears ready to specify and use fabrics for underdrain systems. The background laboratory research was developed by Hoffman and Malasheskie.[7] Forshey[8] reports on a number of projected projects proposed by Pennsylvania. Other states are undoubtedly in similar situations.

9.1.5 Other Published Specifications

It is logical, and indeed to be expected, that the manufacturers and/or users of construction fabrics should be involved in specifying fabric details and properties for specific construction uses. In an area where the "how" precedes the technology of "why" this practical and intuitive insight is invaluable. Some examples follow.

Parks[9] of Monsanto has developed the suggested specifications of Table 9.5. The categories of construction use fall roughly into those of the previous chapters (crack reflection, roads over poor soils, drains, and erosion control, with a separate division in the area of erosion control depending on the size of rip-rap being placed on the fabric liner).

Klassen[10] of Erosion Control Engineering, Ltd., of Toronto, has published a brief overview of the use of fabrics in Canada. Beginning in 1963 with the use of woven polyvinylidene chloride and polypropylene monofilament fabrics, to woven polypropylene fabrics in 1970, to the currently used nonwoven fabrics, he emphasizes that fabric selection criteria should be based on both project-related factors and product-related factors. Stating that there are no Canadian guide specifications or listing of approved fabrics, he presents the performance-related classification shown in Table 9.6.

9.2 Current Research and Development

In order to discuss a recently emerged and rapidly expanding field such as research and development in construction fabrics it is helpful to follow a flow chart. Given in Figure 9.1 is the type of "road map" that should be used in this section. However, as with most large-scale methodologies with wide impact, the R&D effort is fragmented and without any master

TABLE 9.5. Suggested Specifications for Engineering Fabrics in Various Applications

Fabric Property	Type of Construction Use				
	Crack Reflection	Roads Over Poor Support Soils	Drains	Erosion Control[a]	
				−1 ft	+2 ft
Thickness, minimum (mils)	60	60	NA[b]	60	100
Water permeability, (4-in. head) (gal/min/ft^2)	200	200	300	300	200
Grab tensile strength, machine and transverse (lb/in.)	90	90	90	90	200
Grab elongation, machine and transverse (minimum, maximum) (%)	20, 70	15, 70	15, 70	15, 70	15, 60
Trapezoid tear strength, minimum machine and transverse (lb)	50	50	50	50	100
Energy to puncture, minimum (ft-lb × 10^{-2})	NA	4.5	4.5	4.5	9.0
Elongation under 40% continuous load to stability, maximum (hours)	NA	24	NA	24	24
Antiblinding property gradient ratio, maximum	NA	3	3	3	3
Softening point (°F)	325	NA	NA	NA	NA
Filtration efficiency, 100% maximum (μm)	100	100	100	100	100
Mullen burst strength, minimum (psi)	NA	200	200	200	500
Planar water flow/ siphonage, minimum [(qt/hr)/2-in. fabric)	0.5	0.3	0.3	0.3	0.3

Source: Ref. 9.

[a]The −1 ft and the +2 ft relates to the size of the rip-rap, in each case assuming the rip-rap is dropped from a height no greater than 3 ft. The same properties relate to the Corps of Engineers Guide Specifications, where −1 ft equals use without a sand banket. The −1 ft specs are also applicable to use in *silt* fences.

[b]Not applicable.

TABLE 9.6. Classification of Performance Expectations of Synthetic Fabrics[a]

Category	Purpose	Soil Retention (μm)[b]	Water Permeability $[(l/m^2)/sec]$[c]	Tensile Strength (kg/5 cm)	
				Warp	Weft
Soil separation fabrics	To prevent soil of varying characteristics (e.g., granular/organic) from mixing.	50–100	>75	>175	>100
Hydraulic filter fabrics:	To retain soil required to be retained and allow free passage of water under both laminar and turbulent conditions.				
Light duty		50–500	>100	>200	>125
Severe duty		50–500	>150	>200	>150
Stress or load distribution—hydraulic filter fabrics	Inherent in these fabrics, in addition to characteristics of hydraulic filters, is the capability to reduce stress and load factors.	50–500	>150	>1000	>150
Special purpose fabrics	To provide a required combination of the above capabilities to meet specific conditions, including fabrics designed to be used in conjunction with specialized placing techniques.	As dictated by project requirements.			

Source: Ref. 10.

[a] While these values are necessarily arbitrary, they are based on experience gained on many hundreds of projects throughout the world. They are suggested as values which should be used on those projects where failure of the synthetic fabric to meet performance expectations would seriously affect the integrity of a project. Some modification may be required when applied to nonwoven fabrics.

[b] Expressed in microns—soil retention capability depends on both soil and hydraulic conditions (e.g., a mesh opening of 180 μm can effectively contain 60-μm soil particles under certain conditions).

[c] 10-cm column of water.

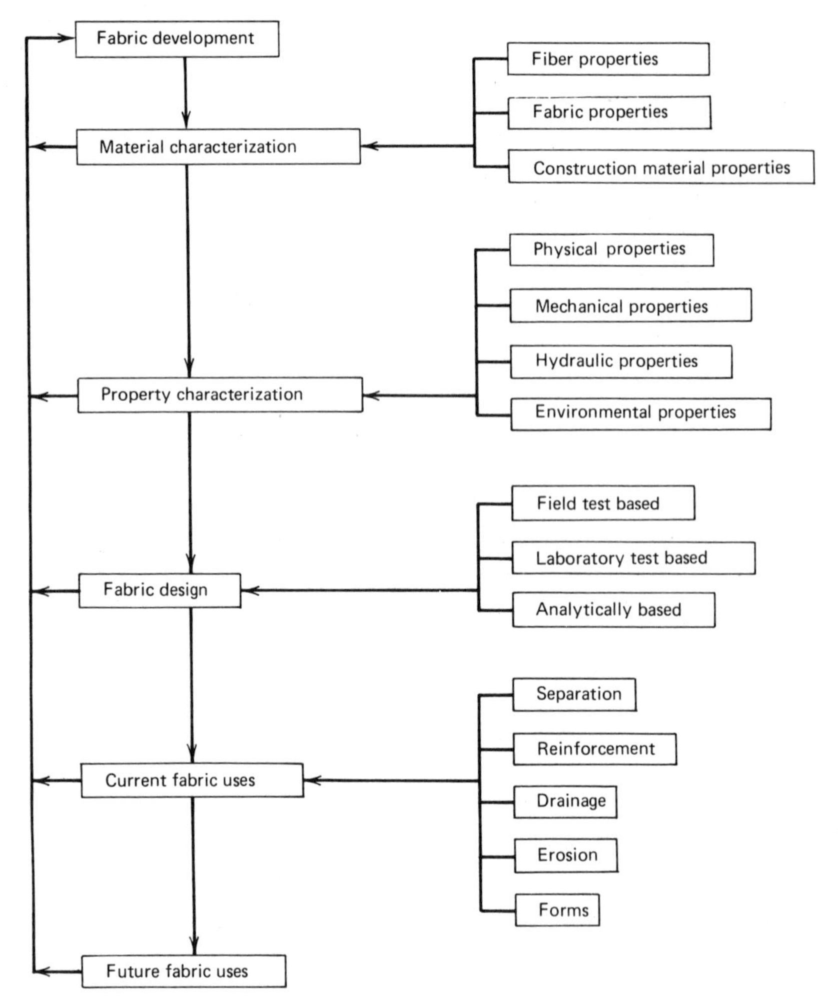

FIGURE 9.1. Flow chart describing idealized research and development efforts in construction fabrics.

plan. Construction fabric R&D is no different. While there is indeed activity, it is difficult to categorize and often results in considerable overlapping of effort.

In this section a brief attempt is made to preview some recent efforts which are considered to be innovative and potentially capable of pushing forward the state of the art. The main headings shown in Figure 9.1 are followed where possible. Obviously the total construction fabric R&D effort cannot be reviewed because of space limitations and a constantly changing scene. Therefore, selectivity on our part is necessary and hopefully it will be typical and representative of the effort currently ongoing.

The area of *fabric development* is probably best left to the chemist or textile engineer; however, it should be realized that the construction and geotechnical engineer is going to be constantly confronted with new and unusual fabrics. As an example, a newly introduced material is Scott polyurethane foam.[11] It differs from the usual nonwoven fabric, originating as a fiber and built into a web, in that this product starts as a chemical mass and ends as a fibrous web. It resembles a nonwoven in structure and is controllable in thickness (from 1/32 to 15 inches), surface area, elongation, filtration capabilities, and permeability. Its performance and general application areas could well include subsurface construction activities.

Regarding *material characterization,* most work is centered around laboratory testing and adapting the many existing test methods for textiles to construction use tests. This is logical since much time and effort has been expended in the past on textile testing. However, new techniques are constantly emerging. An example is Rollins[12] work with an image analyzer to characterize fabric as to the size and shape of the openings and details of clogging or blinding of the fabrics. Crack propogation in fabrics is also being studied on a fracture mechanics basis,[13] indicating that many tools common to other areas are being used in fabric problems.

Finite element methods are being used for *property characterization* and particularly for mechanical property analysis.[14-16] Different elements are used for the soil (which is quite well documented) and for the fabric (which is not well documented). Results are particularly important in indicating the amount of deformation that the fabric will undergo and also information about anchoring, thereby giving an estimate of the required friction between the fabric and the soil. Stress results yield the required tensions that the fabric must sustain. The finite element method should not be considered only an analytical tool since design is also possible. By having a finite element computer code available for a particular problem's geometry, various fabric properties (primarily fabric stiffness) can be evaluated so as to choose the optimum material.

Regarding the hydraulic aspects of *property characterization,* the work of Marks,[17] which has subsequently been extended by Celanese, is of great interest. Filtration mechanisms (and particularly the soil structure arrangement resulting from the water flow), flow data, permeability, and piping are being evaluated and laboratory test methods are being recommended.

The important area of fabric design is in a state of great disarray. Most construction projects seem to be designed on the basis of prior field experiences. Such empiricism does not stand well for the future growth of fabrics, particularly in the public sector.[1] Many people involved in the industry hold high hopes for the laboratory test based methods. But glimpsing back on geotechnical engineering, there is still much scepticism present after 50 years of such methods. (Scale effects will always be present in varying degrees.) The analytic methods, many of which are directly extended from geotechnical engineering, show great promise. To be sure, it is better to work with a calculator in a design office than to foolishly spend large sums trying ill-conceived field programs. The further repercussion of fabric failures in the field is the bad reputation that fabrics may gain if such failures become commonplace.

Regarding *current fabric uses,* the entirety of Chapters 3 through 8 are devoted to the topic. The separate categories listed are taken in order and seem to fit the current patterns of fabric use.

On *future fabric uses,* there appears to be a number of interacting activities that could spur on greater use of fabrics in the future:

- Reports on case histories of field installations with the use of fabrics.
- Capability of being able to specify fabrics in terms of reasonably easy to define properties so that the public sector can be entered.
- Developing appropriate laboratory tests for fabrics.
- Price of the fabric.

9.3 References

1. A. McGown, "The Properties of Non Woven Fabrics Presently Identified as Being Important in Public Works Application," Paper 1, Session 1, Index 78 Programme, University of Strathclyde, Glasgow, Scotland, U.K.
2. *Sample Specifications for Engineering Fabrics,* FHWA-TS-78-211 U.S. Dept. of Transportation, Federal Highway Administration, Washington, D.C. 20590, 1978.
3. C. C. Calhoun, Jr., "Development of Design Criteria and Acceptance Specifications for Plastic Filter Cloths," Technical Report No. S-72-7, U.S. Army Engineers WES, Vicksburg, Miss., June 1972.
4. ______, "Plastic Filter Fabric," Guide Specification No. CW 02215, U.S. Army Engineer WES, Vicksburg, Miss., November 1977.
5. J. Steward, R. Williamson, and J. Mohney, "Guidelines for Use of Fabrics in Construction and Maintenance of Low-Volume Roads," Report No. FHWA-TS-78-205, U.S. Dept. of Transportation, Federal Highway Administration, Washington, D.C., June 1977.
6. L. Lockett, "Use of Filter Fabric in Trench and French Drain Designs," *Highway Focus,* Vol. 9, No. 1, May 1977, pp. 49–62.

7. G. L. Hoffman and G. Malasheskie, "A Laboratory Evaluation of Pertinent Materials and Design Characteristics of PENNDOT's Underdrain System," Pennsylvania DOT Report, December 1977, Harrisburg, Pa.
8. A. D. Forshey, "Use of Filter Cloths for Subsurface Drainage," *Highway Focus*, Vol. 9, No. 1, May 1977.
9. R. M. Parks, "Engineering Fabrics: Their Uses, Evaluation Methods and Specifications," *Highway Focus*, Vol. 9, No. 1, May 1977, pp. 63-71.
10. E. W. Klassen, "Synthetic Fabrics in Civil and Hydraulic Engineering," *Eng. Dig.*, Vol. 22, No. 4.
11. ———, "Scott Polyurethane Foam—Is it a Nonwoven?," *Nonwovens Ind.*, August 1978, pp. 8-13.
12. A. L. Rollin, J. Masounave, and G. Dallaire, "Study of hydraulic behavior of non-woven fabrics," *C. R. Coll. Int. Sols Text.*, Paris, 1977, Vol. II, pp. 201-206.
13. N. J. Abbott and J. Skelton, "Crack Propogation in Woven Fabrics," *J. Coated Fibrous Mater.*, Vol. I, April 1972, pp. 234-252.
14. J. R. Bell, D. R. Greenway, and W. Vischer, "Construction and Analysis of a Fabric Reinforced Low Embankment," *C. R. Coll. Int. Sols Text.*, Vol. 1, 1977, pp. 71-75.
15. E. F. Schwab., O. Pregl, and B. B. Broms, "Deformation Behavior of Reinforced Sand at Model Tests Measured by the X-Ray Technique," *C. R. Coll. Int. Sols Text.*, Vol. 1, 1977, pp. 105-112.
16. R. B. Testa, M. Stubbs, and W. R. Spillers, "A Bilinear Model for Coated Square Fabrics," Report No. NSF ENG 74-18096, Columbia University, New York, July 1977.
17. B. D. Marks, "The Behavior of Aggregate and Fabric Filters in Subgrade Applications," report, University of Tennessee, February 1975.

Appendix 1

Generic Names of Synthetic Fabrics

The U.S. Federal Trade Commission Rules and Regulations under the Textile Products Identification Act names and defines the following generic fabrics.

Pursuant to the provisions of Section 7(c) of the Act, the following generic names for manufactured fibers, together with their respective definitions, are hereby established:

acetate: a manufactured fiber in which the fiber-forming substance is cellulose acetate. Where not less than 92% of the hydroxyl groups are acetylated, the term *triacetate* may be used as a generic description of the fiber.

acrylic: a manufactured fiber in which the fiber-forming substance is any long chain synthetic polymer composed of at least 85% by weight of acrylonitrile units

$$(-CH_2-\underset{\displaystyle CN}{\underset{|}{C}H}-)$$

anidex: a manufactured fiber in which the fiber-forming substance is any long chain synthetic polymer composed of at least 50% by weight of one or more esters of monohydric alcohol and acrylic acid.

$$(CH_2{=}CH-COOH)$$

aramid: a manufactured fiber in which the fiber-forming substance is a long-chain synthetic polyamide in which at least 85% of the amide

$$(-\underset{\displaystyle O}{\underset{\|}{C}}-NH-)$$

linkages are attached directly to two aromatic rings.

azlon: a manufactured fiber in which the fiber-forming substance is composed of any regenerated naturally occurring proteins.

glass: a manufactured fiber in which the fiber-forming substance is glass.

metallic: a manufactured fiber composed of metal, plastic-coated metal, metal-coated plastic, or a core completely covered by metal.

modacrylic: a manufactured fiber in which the fiber-forming substance is any long chain synthetic polymer composed of less than 85% but at least 35% by weight of acrylonitrile units,

$$(-CH_2-\underset{\displaystyle CN}{\underset{|}{CH}}-)$$

except fibers qualifying under subparagraph (2) of paragraph (j) (rubber) of this section and fibers qualifying under paragraph (q) (glass) of this section.

novoloid: a manufactured fiber containing at least 85% by weight of a cross-linked novolac.

nylon: a manufactured fiber in which the fiber-forming substance is a long-chain synthetic polyamide in which less than 85% of the amide

$$(-\underset{\displaystyle O}{\underset{\|}{C}}-NH-)$$

linkages are attached directly to two aromatic rings.

nytril: a manufactured fiber containing at least 85% of a long chain polymer of vinylidene dinitrile

$$(-CH_2-C(CN)_2-)$$

where the vinylidene dinitrile content is not less than every other unit in the polymer chain.

olefin: a manufactured fiber in which the fiber-forming substance is any long chain synthetic polymer composed of at least 85% by weight of ethylene, propylene, or other olefin units, except amorphous (non-crystalline) polyolefins qualifying under category (1) of paragraph (j) (rubber) of rule 7.*

polyester: a manufactured fiber in which the fiber-forming substance is any

*Note should be made that polypropylene, used for many construction fabrics, is an olefin.

long chain synthetic polymer composed of at least 85% by weight of an ester of a substituted aromatic carboxylic acid, including but not restricted to substituted terephthalate units,

$$p(\text{—R—O—}\underset{\underset{O}{\|}}{C}\text{—}C_6H_4\text{—}\underset{\underset{O}{\|}}{C}\text{—O—})$$

and parasubstituted hydroxybenzoate units,

$$p(\text{—R—O—}C_6H_4\text{—}\underset{\underset{O}{\|}}{C}\text{—O—})$$

(As amended September 12, 1973.)

rayon: a manufactured fiber composed of regenerated cellulose, as well as manufactured fibers composed of regenerated cellulose in which substituents have replaced not more than 15% of the hydrogens of the hydroxyl groups.

rubber: a manufactured fiber in which the fiber-forming substance is comprised of natural or synthetic rubber, including the following categories: (1) a manufactured fiber in which the fiber-forming substance is a hydrocarbon such as natural rubber, polyisoprene, polybutadiene, copolymers of dienes and hydrocarbons, or amorphous (non-crystalline) polyolefins, (2) a manufactured fiber in which the fiber-forming substance is a copolymer of acrylonitrile and a diene (such as butadiene) composed of not more than 50% but at least 10% by weight of acrylonitrile units.

$$(\text{—}CH_2\text{—}\underset{\underset{CN}{|}}{CH}\text{—})$$

The term *lastrile* may be used as a generic description for fibers falling within this category, (3) a manufactured fiber in which the fiber-forming substance is a polychlorprene or a copolymer of chloroprene in which at least 35% by weight of the fiber-forming substance is composed of chloroprene units.

$$(\text{—}CH_2\text{—}\underset{\underset{Cl}{|}}{C}\text{=}CH\text{—}CH_2\text{—})$$

saran: a manufactured fiber in which the fiber-forming substance is any long chain synthetic polymer composed of at least 80% by weight of vinylidene chloride units

$$(\text{—}CH_2\text{—}CCl_2\text{—})$$

spandex: a manufactured fiber in which the fiber-forming substance is a long chain synthetic polymer comprised of at least 80% of a segmented polyurethane.

vinyl: a manufactured fiber in which the fiber-forming substance is any long chain synthetic polymer composed of at least 50% by weight of vinyl alcohol units

$$(-CH_2-CHOH-)$$

and in which the total of the vinyl alcohol units and any one or more of the various acetal units is at least 85% by weight of the fiber.

vinyon: a manufactured fiber in which the fiber-forming substance is any long chain synthetic polymer composed of at least 85% by weight of vinyl chloride units.

$$(-CH_2-CHCl-)$$

Appendix 2

List of the Most Common Fiber Trademarks Registered by Member Companies of the Man-Made Fiber Producers Association That Are Currently (1978) Active

Trademark	Generic Name	Member Company
Acrilan	Acrylic, modacrylic	Monsanto Textiles Company
Anso	Nylon	Allied Chemical Corp., Fibers Div.
Antron	Nylon	E. I. du Pont de Nemours & Co., Inc.
Ariloft	Acetate	Eastman Kodak Company, Tennessee Eastman Co. Division
Arnel	Triacetate	Celanese Fibers Marketing Co., Celanese Corp.
Avlin	Polyester	Avtex Fibers Inc.
Avril	Rayon (high wet modulus)	Avtex Fibers Inc.
Bi-Loft	Acrylic	Monsanto Textiles Company
Blue "C"	Nylon, polyester	Monsanto Textiles Company
Cadon	Nylon	Monsanto Textiles Company
Cantrece	Nylon	E. I. du Pont de Nemours & Co., Inc.
Caprolan	Nylon, polyester	Allied Chemical Corp., Fibers Div.
Celanese	Nylon, acetate	Celanese Fibers Marketing Co., Celanese Corp.
Chromspun	Acetate	Eastman Kodak Company, Tennessee Eastman Co. Division
Coloray	Rayon	Courtaulds North America Inc.
Cordura	Nylon	E. I. du Pont de Nemours & Co., Inc.

Trademark	Generic Name	Member Company
Courtaulds nylon	Nylon	Courtaulds North America Inc.
Crepeset	Nylon	America Enka Company
Creslan	Acrylic	American Cyanamid Company
Cumuloft	Nylon	Monsanto Textiles Company
Dacron	Polyester	E. I. du Pont de Nemours & Co., Inc.
Elura	Modacrylic	Monsanto Textiles Company
Encron	Polyester	American Enka Company
Enkaloft	Nylon	American Enka Company
Enkalure	Nylon	American Enka Company
Enkasheer	Nylon	American Enka Company
Enkrome	Rayon	American Enka Company
Estron	Acetate	Eastman Kodak Company, Tennessee Eastman Co., Division
Fibro	Rayon	Courtaulds North America Inc.
Fina	Acrylic	Monsanto Textiles Company
Fortrel	Polyester	Fiber Industries Inc., Marketed by Celanese Fibers Marketing Co., a Division of Celanese Corp.
Herculon	Olefin	Hercules Inc., Fibers Division
Hollofil	Polyester	E. I. du Pont de Nemours & Co., Inc.
Kevlar	Aramid	E. I. du Pont de Nemours & Co., Inc.
Kodel	Polyester	Eastman Kodak Company, Tennessee Eastman Co. Division
Lanese	Acetate, polyester	Celanese Fibers Marketing Co., Celanese Corp.
Loftura	Acetate	Eastman Kodak Company, Tennessee Eastman Co. Division
Lurex	Metallic	Dow Badische Company
Lycra	Spandex	E. I. du Pont Nemours & Co., Inc.
Marvess	Olefin	Phillips Fibers Corp., Subsidiary of Phillips Petroleum Co.
Monvelle	Biconstituent nylon/spandex	Monsanto Textiles Company
Multisheer	Nylon	American Enka Company
Nomex	Aramid	E. I. du Pont de Nemours & Co., Inc.
Orlon	Acrylic	E. I. du Pont de Nemours & Co., Inc.
Polyloom	Olefin	Chevron Chemical Co., Fibers Div.
Qiana	Nylon	E. I. du Pont de Nemours & Co., Inc.
Quintess	Polyester	Phillips Fibers Corp., Subsidiary of Phillips Petroleum Co.
SEF	Modacrylic	Monsanto Textiles Company
Shantura	Polyester	Rohm and Haas Co., Fibers Div.
Shareen	Nylon	Courtaulds North America Inc.

Trademark	Generic Name	Member Company
Spectran	Polyester	Monsanto Textiles Company
Strialine	Polyester	American Enka Company
Teflon	Fluorocarbon	E. I. du Pont de Nemours & Co., Inc.
Trevira	Polyester	Hoeschst Fibers Industries
Twisloc	Polyester	Monsanto Textiles Company
Ulstron	Nylon	Monsanto Textiles Company
Ultron	Nylon	Monsanto Textiles Company
Vecana	Nylon	Chevron Chemical Company
Vectra	Olefin	Vectra Corp., Subsidiary of Chevron Chemical Company
Verel	Modacrylic	Eastman Kodak Company, Tennessee Eastman Co. Division
Zantrel	Rayon (high wet modulus)	American Enka Company
Zeflon	Nylon	Dow Badische Company
Zefran	Acrylic, nylon, polyester	Dow Badische Company

Appendix 3

Trademarks of Nonwoven Fabrics of Member Companies of Man-Made Fiber Producers Association

Trademark	Generic Name	Member Company
Bidim	polyester, spun-bonded	Monsanto Textiles Company
Cerex	nylon, spun-bonded	Monsanto Textiles Company
Duon	olefin, needle-bonded polyester, needle-bonded	Phillips Fibers Corporation, Subsidiary of Phillips Petroleum Company
Enkamat	nylon, melt-bonded	American Enka Company
Mirafi	thermally bonded nylon/polypropylene	Fibers Industries, Inc. Marketed by Celanese Fibers Marketing Company, Division of Celanese Corporation
Petromat	olefin, needle-bonded	Phillips Fibers Corporation, Subsidiary of Phillips Petroleum Company
Reemay	polyester, spun-bonded	E. I. du Pont de Nemours & Company, Inc.
Sontara	spunlaced	E. I. du Pont de Nemours & Company, Inc.
Supac	olefin, needle-bonded	Phillips Fibers Corporation, Subsidiary of Phillips Petroleum Company
Typar	polypropylene, spun-bonded	E. I. du Pont de Nemours & Company, Inc.
Tyvek	olefin, spun-bonded	E. I. du Pont de Nemours & Company, Inc.

Appendix 4

Fabric Organizations

4.1 Trade Organizations

American Textile Manufacturers Institute, Inc.
Economic Information Division
1150 17th Street, N.W.
Washington, D.C. 20036

American Textile Manufacturers Institute, Inc.
1501 Johnston Building
Charlotte, N.C. 28281

Man-Made Fiber Products Association, Inc.
1150 Seventeenth Street, N.W.
Washington, D.C. 20036

Association of the Nonwoven Fabrics Industry (INDA)
100 East 40 Street
New York, N.Y. 10016

Canvas Products Association International (CPAI)
600 Endicott Building
St. Paul, Minn. 55101

Air Structures Institute
(Affiliated with CPAI)
350 Endicott Building
St. Paul, Minn. 55101

4.2 Fabric Manufacturers

American Enka Co.
Enka, N.C. 28728
(704) 667-7110

Celanese Fibers Marketing Company
Box 1414
Charlotte, N.C. 28232
(704) 554-2000

E. I. du Pont de Nemours & Company
Textile Research Laboratory
Wilmington, Del. 19898
(302) 774-0650

Monsanto Textiles Company
800 N. Lindbergh Blvd.
St. Louis, Mo. 63166
(314) 694-7179

Carthage Mills
Erosion Control Division
124 W. 66th Street
Cincinnati, Ohio 45216
(513) 242-2740

Owens-Corning Fiberglas Corp.
Technical Center
P. O. Box 415
Granville, Ohio 43023
(614) 587-0610

PPG Industries, Inc.
One Gateway Center
Pittsburgh, Pa. 15222

Nicolon Corporation
4229 Jeffery Drive
Baton Rouge, La. 70816
(504) 292-3010

GAF Corporation
Glenville Station
Greenwich, Conn. 06830
(203) 324-5418

Menardi-Southern Corp.
3908 Colgate
Houston, Texas 77087
(713) 643-6513

Eastman Chemical Products, Inc.
Kingsport, Tenn. 37662
(212) 262-7187

Bay Mills Midland, Ltd.
Midland, Ontario L4R 4G1
Canada
(705) 526-7867

Phillips Fibers Corp.
P.O. Box 66
Greenville, S.C. 29602
(803) 242-6600

ICI Fibres: "Terram"
Pontypool, Gwent, NP4 8YD
Great Britian
(04955) 58150

Crown Zellerbach
Nonwoven Fabrics Division
Camas, Wash. 98607
(206) 834-5954

Wellington Sears Co.
Marketing Subsidiary of West Point Pepperell, Inc.
111 West 40th Street
New York, N.Y. 10018
(212) 354-9150

Staff Industries, Inc.
78 Dryden Road
Upper Montclair, N.J. 07043
(201) 744-5367

Amoco Fibers Company
Patchague Plymouth Division
550 Interstate North
Atlanta, Ga. 30339
(404) 691-4081

4.3 Fabric Contractors

Soil Stabilization, Inc.
Croton Falls Executive Park
Route 22
P. O. Box 398
Croton Falls, N.Y. 10419

Taiyo Kogyo Co., Ltd.
P. O. Box 831
Redwood City, Calif. 94064
(415) 854-0465

or

Taiyo Kogyo Co., Ltd.
8-4 Kigawahigashi
4-chome, Yodogawaku
Osaka 532, Japan
(06) 305-2111

Erosion Control, Inc.
Forum 111, Suite 507
1655 Palm Beach Lakes Blvd.
West Palm Beach, Fla. 33401
(305) 686-7820

Erosion Control Products, Inc.
1329 Martingrove Rd.
Rexdale, Ontario MOW 4X5
Canada
(416) 745-7290

or

Erosion Control Products, Inc.
Route 5, Box 406
Daphne, Ala. 36526
(205) 626-3510

Advance Construction Specialties Co.
P.O. Box 17212
Memphis, Tenn. 38117
(901) 362-0980

Edward E. Gillen, Co.
218 West Becker Street
Milwaukee, Wis. 53207
(414) 744-9824

Construction Techniques, Inc.
11900 Shaker Blvd.
Cleveland, Ohio 44120
(216) 623-0679

Conenco International Limited
39 Esna Park Drive
Markham, Ontario, Canada L3R1C9
(416) 495-6404

Erosion Control Systems, Inc.
3349 Ridgelake Drive, Suite 101-B
Metairie, La. 70002
(504) 834-5650

VSL Corporation
101 Albright Way
Los Gatos, Calif. 95030
(408) 866-5000

Intrusion Prepakt, Inc.
1705 The Superior Building
Cleveland, Ohio 44114
(216) 623-0080

True Temper Corporation
Railway Appliance Division
1623 Euclid Ave.
Cleveland, Ohio 44115
(216) 696-3366

Hayward Baker Co.
1875 Mayfield Road
Odenton, Md. 21113
(301) 551-8200

Index

AATCC test method 30-1974, 36
Abrasion resistance, fabrics, test for, 30
Acetate, 4, 244, 248, 249
  production of, 5
Acrilan, 248
Acrylic, 244, 248-250
  production, 5
Adriatic Sea, damming of, fabric use in, 223, 224, 227
Adva-Felt, 38
Advance Construction Specialties Co., 38, 255
  Laurel cloth, 44
  Polyfelt, 53
Agricultural drainage, fabric use in, 130
Air permeability test, 31, 32
Airplane cloth, 15
Air Structures Institute, 252
Air-supported structures, fabric use in, 209-217, 219, 221, 222
Akzona Incorporated, Enkamat, 40
Alabama, construction fabric specifications, 233, 234
Aldek, A. S., longard tubes, 171
Alfheim, S. L., 29, 30
Al-Hussaini, M. M., 106
Alkali-resistant glass fibers, 117, 118
Allied Chemical Corp., fiber trademarks, 248
American Cyanamid Company, fiber trademarks, 249
American Enka Company, Enkamat, 40
  fiber trademarks, 249-252
  Stabilenka, 58
American Society for Testing and Materials (ASTM), 26
American Textile Manufacturers Institute, Inc., 252
  textile industry, statistical information, 1, 2
Amoco Fabrics Company, ProPex, 56
Amoco Fibers Company, 254
Anidex, 244
Anso, 248
Antistatic fibers, 7
Antron, 248
Aramid, 244, 249
  production of, 5
Ariloft, 248
Army duck fabric, 15
Arnel, 248
Artificial seaweed, fabric, 175, 178, 180, 181
Asphaltic sealants, 111
Asphalt Institute, flexible pavement system, 89
Association of the Nonwoven Fabrics Industry (INDA), 16, 252
ASTM, construction fabric testing, Committee D-13, 26
    Committee F-17.65, 130
    Subcommittee D-13.61, 26, 130
ASTM test methods, CFMC-FEET-6, 31
  D543, 34
  D737, 31, 32
  D746, 35
  D751, 27-29
  D774, 30, 36
  D794, 35
  D1175, 30
  D1424, 29, 30
  D1435, 35

D1682, 27, 28, 31
D1777, 27
D1910, 26, 27
D2263, 29
D2990, 28, 29
Avisun, artificial seaweed use, 178
Avlin, 248
Avril, 248
Avtex Fibers, Inc., fiber trademarks, 248-250
Azlon, 245

Balloon cloth, 15
Barrett, R. J., 155
Basket weave, 16
Bay Mills fabrics, 39
Bay Mills Midland, Ltd., 254
BC Research, sonic testing methods, 196
Bell, J. R., 21
Belting duck fabric, 15
Benkleman beam test, 110
Benson, G. R., 128
Biaxial tensile test, 28
Bicomponent, 7
Biconstituent, 7
Biconstituent nylon/spandex, 249
Bidim, 39, 40, 251
Bi-Loft, 248
Birdair Structures, Inc., fabric use, 216, 218, 219, 221
Blue "C", 248
Boussinesq theory of stress mobilization, 87-89
Brashears, R. L., 178
Bridge approach roads, construction of, fabric use in, 94
Bridge pier foundations, fabric use in, 157-159
Bridge piers, scour protection, erosion control mat use, 167
underpinning, fabric use in, 190-192
Brittleness, fabric, 35, 36
Broms, B. B., 84, 105
Building materials, plastic fiber reinforcement, use of, 118, 121
Burial deterioration, fabrics, 36
Burlington Industries, Easy-Fencin', 184-185
Burst tearing test, 30
Cadon, 248
Calendering, 16
Calhoun, C. C., Jr., 230
California, fabric specifications, 234
California bearing ratio, 90
California bearing ratio (CBR) mold, 29, 30
Canada, use of fabrics in, 237
Cantilever beam test, 35
Cantrece, 248
Canvas Products Association International (CPAI), 252
air and tension structures, 210, 211, 212, 214
Caprolan, 248
Caricrete, 118
Carthage Mills, 253
Filter-X, 44
Poly-Filter fabrics, 53, 54
Cavern stability, columns for, fabric use, 202-204
CBR, 90
Cedegren, H. R., 124, 126, 133
Celanese Fibers Marketing Corp., Celanese Corp., CFMC FEET-1 test, 31
fabric piping performance tests, 34
fabric research, 241
fiber trademarks, 248, 249, 253
Mirafi, 47
Mirafix 100X, 185
upward gradient tests, 33
water permeability test, 32
Cellulose, 4
Cellulosic fibers, 2, 4
Cerex, 40, 251
CGSB 4-GP-2 method 28.3, 36
Chafer duck fabric, 15
Chemical reagents, fabric resistance to, 34
Chevron Chemical Co., fiber trademarks, 249, 250
Chimney drains, construction of, fabric use in, 134, 135
Chiyoda Chemical Engineering and Construction Co., Ltd., fabric use, 143
Chiyoda Pack Drain Method, 143
Christiani and Nielsen, grout-filled fabric cushion, 194
Chromspun, 248
Coastal structures, liners, fabric use, 155
Colamussi, 223
Colbond KH630 fabric, 144

Cold Regions Laboratory, Corps of Engineers, fabric testing, 149, 150
Coloray, 248
Colorfastness, 7
Combination yarns, forming of, 6
Compressibility, fabric, 27
Concrete, construction forms, fabric patent bibliography, 204-207
  fabric-reinforced, 115-118, 121
  fiber-reinforced, 115-118, 121
  glass fiber use in, 117, 118
  plastic fiber use in, 118
Concrete forms, fabric as, 187-192, 194-196
Concrete piles, land, fabric use, 200-201
  marine environment deterioration, 196, 197
  water, fabric use, 201
Conenco International Limited, 255
  Portadam, 227, 229
  Tubedam, 227
Cone penetration impact test, 30
Consolidation settlement, 141, 142
Construction access roads, fabric use in, 95, 96
Construction fabrics, 2
  details of, 36, 38-41, 44, 46, 47, 49, 52-54, 56, 58-60, 62, 63, 68
  published specifications, 237
  research and development in, 237, 240-242
  uses of, 21-25
Construction forms, concrete and grout, fabric patent bibliography, 204-207
Construction Grade Fibretex, 44
Construction Techniques Incorporated, 255
  Fabridrain, 143
  Fabriform, 163
Containment dikes, dredged materials, fabric use, 99-102
Containment systems, fabric, 106, 107, 109, 110
Continuous heat test, 35
Cordura, 40, 248
Corps of Engineers, bridge approach roads, fabric use in, 94
  Cold Regions Research and Engineering Laboratory, fabric use, 107, 109, 110
    fabric testing, 149, 150
  CW-02215 fabric test, 31, 33
  equivalent opening size test, 31
  erosion control mattress use, 165, 167
  plastic filter cloth testing, 233
  plastic filter cloth use, 232
  seawall fabric liner use, 155
  subdrain guidelines, 230, 231
Cotton, preshrinking, 16
Cotton duck fabric, 15
Count, 16
Courtaulds North America Inc., fiber trademarks, 248, 249
Courtaulds nylon, 249
Creep behavior test, 28, 29
Crepeset, 249
Creslan, 249
Cross dyeing, 7
Crosswise filling yarns, 15
Crown Zellerbach, 254
  Fibretex, 44
Cumuloft, 249
Cyclic heat test, 35

Dacron, 249
Danish Institute of Applied Hydraulics, longard tubes, 171
Dartnell, J. S., 178
Denier, 4, 5
Department of Transportation, Federal Highway Administration, fabric use, 112
  New York State, filter fabric soil retention test, 33, 34
Diaphragm pressure test, 36
Dierickx, W., 130
Dillingham Corporation, fabric use, 198
Dow Badische Company, fiber trademarks, 249, 250
Downdrag, reduction of, fabric use, 201, 202
Drainage, fabric, tests for, 31
  fabric utilization and, 24, 124
Drainage galleries, construction of, fabric use in, 134, 135
Drains, fabric, settlement acceleration, 141-145
Dredged materials, containment dikes, fabric use, 99-102
Drexel University, upward gradient tests, 33
Dry spinning, 5
Dunham, J. W., 155

Duon, 251
du Pont, 29
  chemical reagents tests, 34
  Cordura, 40
  fabric testing, 90
  fiber trademarks, 248-251, 253
  light and weather resistance tests, 35
  Reemay, 56
  Sontara, 56
  Typar, 63, 121
  Tyvek, 63, 68
  water permeability test, 32
Dura-Bags, 169
Dye nets, 16

Earth dams, fabric use in, 106, 133-137
Earth walls, fabric reinforced, 105, 106
Eastman Chemical Products, Inc., 253
Eastman Kodak Company, fiber trademarks, 248-250
Easy-Fencin', 184, 185
Eaton, R. A., 110
Edward E. Gillen, Co., 255
  Longard Tube System, 171
E. I. du Pont de Nemours & Co., Inc., *see* du Pont
Einsenmann, J., 80
Elmendorf tear test, 29, 30
Elongation tests, 28
Elura, 249
Embankments, soft soils, fabric use in, 95
Encron, 249
"Ends", 15
Endurance tests, fabrics, 34
Enka, water-filled structures, fabric use in, 223
Enka b.v., fabric use, 157-159
Enkadrain, 40, 41, 44, 139, 141
Enkaloft, 249
Enkalure, 249
Enkamat, 40, 41, 44, 251
Enkamat Matting, 40
Enkasheer, 249
Enkrome, 249
ENPC, 21
EOS, 31
Equivalent opening size (EOS) test, 31
Erosion control, fabric use in, 24, 25, 149-151
Erosion Control, Inc., 255
  Dura-Bags, 169
  Easy-Fencin', 184, 185
  Silt-Ban, 184
Erosion Control Engineering, Ltd., fabric specifications, 237
Erosion control mattresses, 160, 162-165, 167-169, 171
Erosion Control Products, Inc., 255
Erosion control structures, construction fabric use, 152
Erosion control systems, fabric installation, 153, 154
  fabric liners, 151-160
Erosion Control Systems, Inc., 255
  Gobimat, 168
Erosion prevention, fabric use in, 147
Estron, 249

Fabricast molded blocks, 169, 171
Fabric construction, elements of, 13, 15-21
Fabric containment systems, 106, 107, 109, 110
Fabric contractors, 254-256
Fabric design, developments in, 241, 242
Fabric development, 241
Fabric drainage, tests for, 31
Fabric drains, settlement acceleration, 141
Fabric earth dams, 106
Fabric-enclosed sand drains, 143
Fabric finishing, 16
Fabric forms, pile jacketing, 196-202
Fabric interceptor system, 130, 131
Fabric manufacturers, 252-254
Fabric organizations, 252-256
Fabric piping performance tests, 34
Fabric properties, construction use, importance of, 26-36
Fabric-reinforced concrete, 115-118, 121
Fabric retaining walls, complete, 103-105
Fabrics, construction applications, selection of, 26-36
  endurance tests, 34-36
  hydraulic properties, tests for, 31-34
  mechanical properties, tests for, 27-31
  miscellaneous properties tests, 34-36
  opening size, tests for, 31
  physical properties, tests for, 26, 27
  porosity, tests for, 31
Fabric silt fencing, 181, 184, 185
Fabric thickness, 27

Fabric trade organizations, 252
Fabric use, guidelines for, 230-234, 236, 237
Fabric uses, future, 242
Fabric weight, 26, 27
Fabridrain, 143
Fabriform, 163
Fatigue resistance, fabrics, test for, 31
Federal Environmental Protection Agency, 181
Federal Highway Administration, fabric use, 112
Federal Standard No. 191, method 5762, 36
Federal Trade Commission, generic fiber definitions, 8
Fiber base, 4, 5
Fiberfill, 6
Fiberglass, Ltd., alkali-resistant glass fibers, 117, 118
Fiber Industries Inc., fiber trademarks, 249, 250
Fibermarkers, Ltd., Terrafirma, 168
Fiber production, synthetic, 4-8
Fiber-reinforced concrete, 115-118, 121
Fiber statistics, synthetic, 1, 2
Fiber trademarks, 248-250
Fiber variants, 8, 12
Fibretex, 44
Fibro, 249
Filter duck fabric, 15
Filter fabric soil retention test, 33, 34
Filter Point erosion control mat, 163
Filter-X, 44
Filtram, 60, 62, 63, 141
Fina, 249
Fine-combed cotton fabric, 15
Fine-grained soil, containment of, fabric use in, 107, 109, 110
Finishing, fabrics, 16
Finnigan, J. A., 29
Flame retardant fibers, 7
"Flat" duck fabric, 15
Fluorocarbon, 250
Foreshore Protection Property, Ltd., Terrafirma, 168
Forms, fabric materials and, 25
Forshey, A. D., 237
Fortrel, 249
Frankipile, Ltd., pile protection, 201
French drain system, fabric use in, 126, 127
"Frost boils", 115
Fulanis, fabric use, 223

GAF Corporation, 253
Gated storm-surge barrier, fabric use in, 192, 194
Generic fibers, 8
Geotechnical construction, fabrics used in, 36, 38
Gernan Federal Railway, fabric use in, 80
Giroud, J. P., 136
Glass, production of, 5
Glass fibers, 115, 245
applications of, 118
in concrete, 117, 118
Gobibricks, 168
Gobimat, 168, 169
Grab tensile test, 28, 31
Gradient ratio test, 33
Gray, R. E., 202
Grout, construction forms, fabric patent bibliography, 204-207
Gour-filled fabric cushion, 194
Guide Specifications Plastic Filter Cloth, 232
Gulf States Paper Company, erosion control fabrics, 149

Haliburton, T. A., 100
Handbook of Industrial Textiles, 8
Hayward Baker Co., 256
Healy, K. A., 130, 131
Healy and Long system, drainage, fabric use in, 130, 131
"Heat setting", 16
Hercules, Inc., fiber trademarks, 249
Herculon, 249
Heterofilaments, 20
Heterofil bonding, 20
Hicks, A. B., 21
Highway Research Board, scour depths, 157
Highway underdrains, fabric use in, 128
Hoedt, G. den, 157, 192
Hoechst Fibers Industries, fiber trademarks, 250
Hoff, G. C., 116, 118
Hoffman, G. L., 237
Hollofil, 249
Homofilaments, 20

Homofil bonding, 19, 20
Hoogendoorn, J., 155
Hose duck fabric, 15
Hydraulic properties, fabrics, tests for, 31
Hydro-Lining Concrete mat, 167, 168
Hydrometer analysis, 34

ICI Fibers, Filtram, 60, 62, 63
  Terram, 60, 62, 63, 254
ICI Linear Composites Scour Prevention System, 159, 160
Illinois, fabric specifications, 234, 236
Illinois Department of Transportation, fabric use, 128
"Image analyzers", 31
Impact strength test, 29, 30
Imperial Chemical Industries, Ltd., Filtram, 60, 62, 63
  Terram, 60, 62, 63, 254
Impermeable fabrics, 25, 209
INDA, 16, 17
Indiana and Michigan Electric Co., electric generating plant roads, fabric use in, 95, 96
Indirect tensile test, 29
In situ soil, 89
International Conference of Coastal Engineering, artificial seaweed testing, 178, 181
International Conference on the Use of Fabrics in Geotechnics (Soils), 21
Interstate 24, Johnson County, fabric use in, 129
Interstate 55, Chicago, fabric use in, 129, 130
Interstate 64, St. Clair County, fabric use, 128, 129
Interstate 95, Pittsfield, Maine, fabric use, 112, 113
"In the gray", 16
Intrusion-Prepakt, Inc., 256
  Fabricast, 169, 171
IT impregnated textile, 83

Johnstown, Pennsylvania, erosion control mattress use, 163-165
J. P. Stevens Co., Inc., Monofelt, 47
  Monofilter fabric, 47, 49

Karim, M., 167
Kaswell, E. R., 8
Kern, F., 106
Kevlar, 249
Kjellman, W., 143
Klassen, E. W., 237
Knit fabric, 13
Kodel, 249

Lacroix, Y., 72, 77
Lanese, 249
Lastrile, 246
Laundry nets, 16
Laurel Cloth, 46
Laurel Erosion Control Cloth (LECC), 46
Laurel Plastics Co., Laurel Cloth, 46
La Verne College, facric use, 219, 221
Lawn fabrics, 15
LCPC, 21
LECC, 46
Lengthwise yarns, 15
Leno weave, 16
Lewis, M., 21
Leykauf, G., 80
Light, fabric resistance to, 35
Lightweight fill, containment of, fabric use in, 107
Linear Composites, Ltd., artificial seaweed use, 180
LNG offshore facility, fabric tubes for, 194-196
Loading, terminal bases, fabric use in, 159
Lockett, L., 234
Loftura, 249
Long, R. P., 130, 131
Longard tubes, 171
Longard Tube System, 171
Loudier, D., 136
Lurex, 249
Lycra, 249

McGown, A., 17, 230
McGuffey, V. C., 139
Maine Department of Transportation, fabric use, 115, 154
Malasheskie, G., 237
Mandrel, 142
Man-Made Fiber Producers Association, Inc., 252
  common trademarks, 12
  fiber trademarks, 248-250

nonwoven fabric trademarks, 251
synthetic fiber production statistics, 2
Manning formula, 126
Marks, B. D., 241
Marvess, 249
Material characterization, developments in, 241
Materials, separation of, fabric use in, 71
Mechanical properties, fabrics, tests for, 27-31
Melt-bonded process, 19, 20
Melt spinning, 5
Menardi-Southern Corp., 253
Monofelt, 47
Monofilter fabric, 47, 49
Mesh nets, synthetic, erosion control use, 150, 151
Metallic fiber, 245, 249
Miller, S. P., 147
Mine stability, columns for, fabric use, 202-204
Mirafi 100X, 47, 185
Mirafi 140 fabric, 47
Mirafi 500X, 47
Miscellaneous properties, fabrics, 34-36
Mission Dam, British Columbia, Canada, fabric use in, 72, 77, 188-190
Mitchell, J. K., 97
Modacrylic, 245, 249, 250
production of, 5
Monofelt, 47
Monofilament yarn, 7, 8
Monofilter fabric, 47, 49
Monsanto Textiles Company, Bidim, 39
Cerex, 40
fabric creep behavior test, 29
fabric specifications, 237
fiber trademarks, 248-251, 253
Monvelle, 249
Mouw, K. A. G., 157, 192
Mullen burst test, 30, 36
Multifilament yarn, 7, 8
Multisheer, 249
Muskeg, road construction, fabric use, 95

Napping, 16
NASA, artificial seaweed testing, 178
National Experimental and Evaluation Program (NEEP), fabric use, 112
National Research Council of Canada, burial deterioration tests, 36
Natural fibers, 1
Needle punched fabric, production of, 18, 19
Needle punched process, 18, 19
Negative skin friction, fabric use in, 201, 202
Netherlands State Road Laboratory, fabric tests, 95
New Jersey Turnpike Authority, fabric use, 200, 201
New York State Department of Transportation, fabric specifications, 236, 237
fabric use, 139
filter fabric soil retention test, 33, 34
Nicolon artificial seaweed, 178, 180
Nicolon Corporation, 253
Nicolon fabric, 49, 139
Nicolon polypropylene seaweed, 178, 180
Nomex, 249
Noncellulosic fibers, 2
production of, 4
Nonwoven fabrics, 2, 13, 16-21
manufacturing system, 17
trademarks, 251
Novoloid, 245
"Numbered" duck fabric, 15
Nylon, 2, 12, 245, 248-251
production of, 5
Nytril, 245

Occidental of Britain, artificial seaweed use, 180
Octagonal fibers, 7
Offshore drilling platforms, fabric use in, 159
Olefern, 178
Olefin, 2, 12, 245, 249-251
production of, 5
Olympic National Forest Wall, fabric retaining wall use, 102, 103
Opening size, fabric, tests for, 31
Oregon State University, construction fabric uses, 21
Orlon, 249
"Outdoor Weathering of Plastics", 35
Outlet drains, construction of, fabric use in, 135, 136
Owens-Corning Fiberglas Corp., 253

air-supported structure, fabric use, 216, 218
alkali-resistant glass fiber, 117, 118
erosion control fabrics, 149

Parks, R. M., 32, 237
Patent bibliography, fabric use, concrete and grout construction forms, 204-207
Pavement deterioration, fabric use in, 114, 115
Pavement underdrain systems, fabric use in, 124-131
Pazsint, D. A., 107
Peck, R. B., 102
Pennsylvania, fabric specifications, 237
Pennsylvania Department of Transportation, fabric use, 190-192
Pentagonal fibers, 7
Perforated pipe system, fabric use in, 126, 127
Perforated sheets, erosion control use, 150, 151
Permealiner fabric, 52, 53
Petromat, 53, 251
Philips Fibers Corp., fiber trademarks, 249, 251, 254
Petromat, 53
Supac, 58, 59
Philips Petroleum Company, Petromat, 53
Supac, 58, 59
Physical properties, fabrics, tests for, 26, 27
"Picks", 15
Pier 16 of L.R. 25, Northumberland County, fabric use, 190-192
Pile jacketing, fabric forms for, 196-202
Pile protection, chemical attack, fabric use, 201
Piles, rehabilitation of, fabric use, 197-199
Pins, securing, plastic filter cloth, 233
Pirelli, 223
Plain weave fabrics, 15
Planar water flow test, 33
Plane strain tensile test, 28
Plassar ballast undercutter cleaner, 81
Plastic fiber reinforcement, building materials, use in, 118, 121
Plastic fibers, 115, 116
Plastic filter cloth, 231, 232
Plastics, high-temperature testing, 35
Polyester, 2, 12, 245, 246, 248-251
production of, 5
Polyfelt, 53
Poly-Filter GB, 53, 54
Poly-Filter X, 53, 54
Polyloom, 249
Polymers, 7
Polypropylene, 12, 251
Polypropylene foam seaweed filament, 181
Pontiac Stadium, fabric use in, 216, 218
Porosity, fabric tests for, 31
Portland cement concrete, 117
Portland cement concrete pavements, cracked, 112
Portadam, 227, 229
Portomod, 221
PPG Industries, Inc., 253
Property characterization, developments in, 241
Preconstruction fill, 142
Predipping, 16
Preshrinking fabrics, 16
ProPex, 56

Qiana, 249
Quantitative metallography, 31
Quintess, 249

Railroad ballast/subgrade separation, fabric use in, 80, 81, 83
Railway Track and Structures, 81
Rainfall, erosion of soil, 148, 149
Rankine earth pressure theory, 102
Rayon, 4, 246, 248-250
production of, 5
Reemay, 56, 251
Reflective cracking, control of, fabric use in, 110-115
Reinforcement, fabrics and, 23, 24, 84, 89, 94-98
Resin bonding process, 20
"Resistance of Plastics to Chemical Reagents", 34
Retaining walls, fabric use in, 102-106, 137-139, 141
Risseeuw, P., 144
Road base, fabric use in, 107, 109, 110
Road construction, fabric use in, 87-90, 92-97

Road reinforcement, construction fabrics, use of, 89, 90, 92
Road subgrade reinforcement, fabric use, placement procedure, 92-94
Rohm and Haas Co., fiber trademarks, 249
Rollins, A. L., 241
Route 118, Waterford, Maine, fabric use, 154
Route 150, Parkman, Maine, fabric use, 115
Royal Dutch Shell Plastics Laboratory, artificial seaweed use, 181
Rubber reinforcement, 16
Rubble revetment, fabric use in, 155

Sand drains, construction of, fabric use in, 142-145
Sand-filled steel pipe, 142
Saran, 246
Sateen, 16
Satin weave fabric, 16
Scour protection, bridge piers, erosion control mats, 167
Seaman Building Systems, tension structures, 221, 222
Seaweed, artificial fabric, 175, 178, 180, 181
Securing pins, plastic filter cloth, 233
SEF, 249
Self-sustaining structures, fabric use in, 227, 229
Separation, fabrics and, 22, 23
  materials, fabric use in, 71
Settlement acceleration, fabric drains, 141-145
Shallow foundations, bearing capacity, fabric use in, 96, 97
Shantura, 249
Shareen, 249
Sheeting fabric, 15
Shell Chemical Company, Caricrete, 118
Sherard, J. L., 133
Short fibers, 6
Silt-Ban, 184
Silt fencing, fabric, 181, 184, 185
Singeing, 16
Siskiyou National Forest, fabric retaining wall use, 104
Sissons, C. R., 28
Skanska Cementgjuriet, grout cushion use, 194
Skelton, J., 212, 213
Slope instability, fabric reinforcement against, 99
Slope stability, fabric use in, 98-102
Smith, N., 107
Soft soils, embankments over, fabric use in, 95
Soil erosion, rainfall and, 148, 149
Soil loss, reducing fabric use in, 149, 150
Soil-retaining mechanism, fabric as, 148-151
Soil Stabilization Inc., 254
Solution dyeing, 7
Sontara, 56, 251
Sorlie, A., 29, 30
South Carolina Highway Department, fabric use, 89
Spandex, 246, 249
  production of, 5
Spectran, 250
Spinneret, 4, 5
Spinning, 5
Spinning systems, 6
Spun-bonded process, 19
Spunlaced, 251
Spun yarns, 6, 7
Stabilenka, 40, 58
Staff Industries, Inc., 254
  Permealiner fabric, 52, 53
Staple, 6, 8
Staple fibers, production of, 5, 6
State of the Art of Subsidence Control, 202
State Department of Transportation, fabric specifications, 234, 236, 237
State University of New York, Buffalo, fabric use, 218, 219
Steel fiber concrete, 117
Steel fibers, 115
Steel piles, land, fabric use, 201
  marine environment deterioration, 197
Steel pipe, sand-filled, 142
Stevenson Expressway, Chicago, fabric use in, 129, 130
Steward, J., 231
Stone seawall, fabric use in, 155-157
Streets, Kinloch, Missouri, fabric use in, 114, 115
Strialine, 250
Strip tensile test, 27
Subdrains, guidelines, Corps of Engineers, 230, 231

Sun Oil Co., artificial seaweed use, 178
Supac, 251
Surcharge loading, 142
Synthetic fabrics, generic names, 244-247
Synthetic fibers, 1, 2
  blending, 7
  forms of, 7, 8
  generic names, 8
  physical properties of, 12
  production of, 4-8
  spinning, 5
Synthetic fiber yarns, texturing, 6
Synthetic mesh nets, erosion control use, 150, 151
Syphonage ability, fabric, 33
Syphonage test, 33

Taiyo Kogyo Co., Ltd., 255
Teflon, 250
Temperature, fabric resistance to, 35, 36
Tension Span Structures, 221
Tension structures, fabric use in, 209, 217, 219, 221, 222
Terrafirma, 168
Terrafix, 169
Terrafix Erosion Control Products, Inc., Terrafix, 169
  Terrafix filter mats, 59
Terrafix filter mats, 59, 60
Terram, 60, 62, 63
Terrapakt fabric forms, 168
Terzaghi, K., 72, 77, 102, 124, 141
Tex, 5
Texas Department of Highways, fabric testing, 113, 114
Textile engineer, fabric construction and, 13
Textile industry, statistical information, 1, 2
Textile Products Identification Act, synthetic fabric generic names, 244-247
Thickness, fabric, 27
Tingstad Tunnel, Gotherburg, Sweden, fabric use in, 194
Toe drains, construction of, fabric use in, 134, 135
Tongass National Forest, fabric tests, 95
Tongue tear test, 29
Tow, 5, 6, 8
Trademark, 12
Trade organizations, fabric, 252
Transition zones, construction of, fabric use in, 134, 135
Transverse elasticity test, 29
Trapezoidal tear test, 29
Trevira, 250
Triacetate, 4, 244, 248
  production of, 5
Trilobal fibers, 7
True Temper Corporation, 256
  material separation, fabrics for, 81, 83
Tubedam, 227
Twill weave fabric, 15
Twisloc, 250
Typar, 63, 121, 251
Typewriter ribbon fabric, 15
Tyvek, 63, 68, 251

Ulstron, 250
Ultron, 250
Underpinning bridge piers, fabric use in, 190-192
Underpinning caissons, fabric use, 192, 194
Underwater tunnel, cushion for, fabric use, 194
Uniform Cross Section erosion control mat, 163
United States Department of Agriculture, Forest Service Specification, fabric use, 231-233
United States Textures Sales Corporation, Nicolon fabric, 49
University of Bologna, fabric use, 223
University of Riyadm, fabric use, 219
Upward gradient test, 33
U.S. Army Corps of Engineers, plastic filter cloth use, 232
U.S. Army Engineer District, Mobile, Alabama, fabric tests, 100-102
U.S. Army Engineers Waterways Experiment Station, fabric tests, 105, 106
U.S. Federal Trade Commission Rules and Regulations, synthetic fabric generic names, 244-247
U.S. Forest Service, fabric use, guidelines and specifications, 231-233
U.S. Government Printing Office, 231
U.S. National Forests, fabric retaining wall use, 103
U.S. Standard sieve, 31

Valcros Dam, fabric use in, 136
Van den Elzen, L. W. A., 144
Variants, 8
Vecana, 250
Vectra, 250
Vectra Corp., fiber trademarks, 250
Verel, 250
Vermont Soil Conservation grass seed mixture, 150
Vinyl, 247
Vinyon, 247
  production of, 5
VSL Corporation, 255
VSL Hydro-Lining, 167, 168

Water-filled structures, fabric use in, 222-224, 227
Water permeability test, 32
Water Quality Considerations for Construction and Dredging Operations, 181
Water Quality Office, fabric silt fencing, 181
Waterways Experiment Station, fabric tests, 94
Weather, fabric resistance to, 35
Weave, 15
Weight, fabric, 26, 27
Wellington Sears Co., 254
Wet spinning, 5
Wood piles, marine environment deterioration, 196
Woven fabrics, 2
  construction of, 13, 15, 16
Woven plastic filter fabrics, U.S. Forest Service specifications, 231
Wrapped underdrain system, 130, 131

Zantrel, 250
Zeflon, 250
Zefran, 250
Zoned earth dams, material separation, fabric use in, 72, 77